PERSONAL EMPOWERMENT

ACHIEVING INDIVIDUAL &
DEPARTMENTAL EXCELLENCE

PERSONAL EMPOWERMENT

ACHIEVING INDIVIDUAL & DEPARTMENTAL EXCELLENCE

BENNIE L. CRANE & DR. JULIAN L. WILLIAMS

PennWell Corporation
1421 South Sheridan Road
Tulsa, Oklahoma 74112

800-752-9764
sales@pennwell.com
www.pennwell-store.com
www.pennwell.com

cover design by Clark Bell
book design by Robin Remaley

Library of Congress Cataloging-in-Publication Data

Crane, Bennie L., 1935-
Personal empowerment : achieving individual and departmental excellence / Bennie L. Crane with Julian L. Williams; with a foreword by Mark Allen Davis.
p. cm.
Includes index
ISBN 0-87814-842-6
1.Fire fighters--Training of. I. Williams, Julian L. II. Title.

TH9120 .C73 2002
363.37'023--dc21 2002066311

Printed in the United States of America

1 2 3 4 5 06 05 04 03 02

Table of Contents

Foreword

Personal Empowerment challenges each of us to confront the enigma that has plagued the cognizance of both race and gender conflicts. Bennie L. Crane demonstrates that the core of our existence lies in the psyche, and that despite our differences we will find a common ground in our humanity. Valuing our cultural differences is not just the right thing to do, it is the substance of a democratic society and essential to the continued growth and prosperity of a nation. *Personal Empowerment* provides an edified corridor for our future.

Each chapter provides an extended understanding of how we function as human beings and of the responses we create toward diverse cultural and gender encounters, which we instinctively address but seldom analyze.

Differences in people have often precipitated conflicts. Social scientists have made extraordinary strides in identifying and labeling human behavior; however, we must understand that "different" does not mean "bad". As human beings we must be willing to accept the differences yet still protect our own interests. When we can demonstrate all of this potential and successfully communicate this message, we can bridge the gaps between cultural and gender heterogeneity and prepare for an indisputable positive outcome.

Bennie L. Crane has incorporated this strategy into his book, and one of the conundrums that has long separated the races is here for your examination.

Mark Allen Davis, M.A.
District Commander
Chicago Police Department

Preface

THIS BOOK IS DEDICATED to honor the victims, survivors, and rescuers of the September 11, 2001 tragedies in New York City and Washington, DC. It is our hope that this presentation of our experiences on that day and the days that followed, will in some way let you know that an entire nation shared in your sorrow and your grief as members of your extended family. Although I live in Chicago, my co-author, Dr. Julian L. Williams, lives in Manhattan. The first chapter, entitled "Let the Healing Begin", will give you our impressions from those two locations. We hope that each of you will find some comfort in this presentation and some additional strength to carry on with your lives. A great and grateful nation is indebted to you. May the good Lord bless each of you.

Chapter 2 is about creating a personal agenda. We suggest that you consider a personal agenda as a means for living your life in a manner that will be an ongoing tribute to your deceased loved ones. Long after the visitors are gone, you have to continue on. Your personal agenda can serve as your touchstone, providing the inspiration that can only come from a plan created by you for you. You are encouraged to review this section and use the easy-to-follow instructions. Additional support material can be found in the "Personal Motivational Workbook" (Appendix A of this text).

That life is a book worth reading, a race worth running, and a song worth singing is what this book is all about. One of the major questions that I have had about the human condition is: "What tools do we have to address the many challenges that we face as we move through life?" I always thought that if I had that information I would be able to work my way through life's many challenges. I am grateful to Stephen R. Covey for providing the answer to this question by identifying endowments that are unique to human beings. In his best selling book, *The 7 Habits of Highly Effective People* (Simon & Schuster), he identifies the endowments that, for me, provide a common ground for humanity. This information had a kind of liberating effect on me. It became clear to me that each human being—no matter what their race or ethnicity, whether they are rich or poor, male or female—comes into the world with freedom of choice, self-awareness, imagination, and an independent will. To the extent that we acknowledge and use these gifts, we will be effective in reducing the effects of stress in our everyday living. This will increase the amount of personal happiness and joy in our lives. Please read on as we put this into perspective for you.

Have a wonderful rest of your life.

Bennie L. Crane
District Chief (Ret)

Julian L. Williams, Ph.D.

Acknowledgements

WE WOULD LIKE TO EXPRESS OUR APPRECIATION to Julian's aunt and my wife Lois, our daughters Linda and Shirley, and their husbands William and Benjamin, for their unwavering moral support and encouragement.

We are very grateful to our publisher Margaret Shake, supervising editor Jared Wicklund, and their staff at Fire Engineering Books & Videos for their technical expertise and their fine job of putting the finishing touches on the book.

We want to recognize and say thank you to the members of our personal support group that were always available to lend us a hand when we needed them. They are Leslie Noy, Richard Kolomay, Robert Hoff, Cynthia Onore, Mark Davis, and Shemicka White.

I would be remiss should I fail to mention the value of the opportunities afforded me as a member of the Chicago Fire Department and the Fire Service Institute of the University of Illinois. The wealth of experience and the many relationships that I developed at these fine organizations contributed greatly to the contents of this book.

Introduction

ONE OF THE MOST FREQUENTLY DISCUSSED socio-cultural issues still prevalent today is the psychological mindset known as "Black rage/White guilt". Stemming from the horrors of the Black enslavement, this rage/guilt is as difficult to some to come to terms with as any debilitating disorder. When juxtaposed with the reverse mode of thinking, "White rage/Black guilt" often times addresses enforced hiring quotas as a result of Affirmative Action legislation. This line of discussion can also be applied to gender tensions in the workplace. Traditionally, men have been paid more money and have received advancements and higher ranks for performing the same jobs as women. This imbalance provides the basis for "Female rage/male guilt". Just as controversial, as a result of Equal Employment Opportunity laws passed in 1964, women are competing for jobs that have been traditionally filled by men (such as heavy equipment operators, construction laborers, and firefighters). This additional competition has fostered what can be termed "Male rage/female guilt".

This process of shifting the advantage from one group to another continues the rage/guilt cycle. With these factors in mind, our objective is to reduce the friction between opposing groups in our society and de-emphasize the divisions between people, focusing less on what we do not have in common and more on what we do have in common. Furthermore, this work is my attempt to put forth the notion that there is a human element often overlooked in the struggle for racial and gender equality. Many of the efforts that address diversity in our society focus on the differences among us. I suggest, however, that if we shift the focus to the similarities among us—to the qualities we all share, even in a diverse society—we will find common ground. Essentially, no matter what race, gender, or other sociocultural group we are born into, we all have one thing in common: our humanity. From this common ground we can begin to establish and work toward common goals.

Let me explain how this understanding affected my professional life. In November 1979, I was promoted to the rank of lieutenant in the Chicago Fire Department. I realized then that my new responsibilities called for broadening my point of view; I could no longer look at issues only from the perspective of my own race and gender. It became necessary for me to relate to the race and gender concerns of each of my subordinates. To do otherwise would have resulted in me displaying the types of behavior I had disliked in some of the officers that I had previously served under.

For a number of reasons, I had never expected to be promoted, not only because I am Black, but also because I had no political connections. In Chicago at that time, political connections generally served as one's most valuable asset. Nonetheless, contrary to my expectations, I did receive the promotion, and I wanted my people to be the best that they could be—for their safety and for their personal pride. My own personal agenda required that I develop the best fire company possible. My experience indicated that firefighters who knew their jobs well had fewer problems than those who did not. It was during this time in my career that I truly began to understand that the color of a person's skin does not determine his or her value or worth. For some reason, this concept had been unclear to me until then, the result of past conditioning I suppose. Leading and training members from all races soon put an end to my stereotypical thinking. The fact was that I now had a responsibility to all the people who worked for me as well as those I worked for, not to mention to the public that we all served. Measuring up to these challenges left no room for unproductive thinking. I needed to learn quickly how to develop new and positive ways of evaluating situations and generating ideas.

Fighting fires is very dangerous work, and fire is an unforgiving foe. Any mistakes in our efforts to control a fire can be, and often are, fatal. No fire officer wants to tell the children or other family members of a deceased firefighter that their loved one is not coming home. Training is one of the most important means of ensuring that firefighters' actions on the fireground are safe and effective.

In September 1982, I took on additional training responsibilities as the district training officer for the 25 fire companies assigned to Chicago's Sixth District. I also served as the field instructor for the University of Illinois Fire Service Institute, which required me to train firefighters working in fire departments throughout the state of Illinois. I discovered that the common thread of my success in all these training activities was the fact that I related to each person first and foremost as a human being. The concepts that I advocate here served me well both as a training officer and later as a district chief with the Chicago Fire Department.

In October 1985, I received the unprecedented appointment from lieutenant to assistant director of training. As a result of the 1980 Firefighter Strike, there were many problems plaguing the department at that time. Members of families were not speaking to one another because some had chosen to work during the strike while others had not. The Labor Relations division was receiving 100 grievances a month. It was obvious that high-quality training would be difficult at best in that atmosphere. So, after gathering information from the National Fire Academy and the University of Illinois, I developed and delivered a presentation on human behavior and conflict resolution to 4,200 department members in 6 weeks. The results were immediate; grievances were reduced and still remain at fewer than 50 per month. The relationship between the training division and the fire suppression companies improved greatly. In the past, attracting qualified, experienced officers to the training academy as instructors had been a problem, but as a result of the new emphasis on conflict resolution, officers began to submit requests to join our staff, reflecting the increased level of interest and confidence in the training division.

The next three years would bring about some significant changes within the Chicago Fire Department. An ambulance assistance program that required fire suppression companies to respond to emergency medical calls was put in place. To be prepared to handle a medical crisis, our suppression people would have to undergo intense training in emergency medical care. Resistance ran high. This was because many firefighters feared that they might do more harm than good, and they were initially opposed to expanding the scope of their jobs to include responding to medical emergencies.

Also, for the first time ever, six female firefighter candidates were included in a recruit class. Initially, this change caused an uproar throughout the department because of gender friction as well as the perception that firefighting is a man's job. The female candidates had all participated in a pre-hire program designed to build their upper-body strength. This enabled the women to be on par physically with the male recruits. In some ways, the presence of the female recruits brought out the best in the male recruits. Some of the men did not want to risk being outperformed by females. Because they were all subjected to the same rigorous training standards, strong bonds naturally developed between the male and female recruits.

In March of 1988, I requested and was granted a field assignment to the Fourth District as a deputy district chief. My responsibilities included directing and supervising the activities of 200 members assigned to the Third Platoon. My goals were to establish and maintain a high level of morale so that the platoon members would support one another and work together as a cohesive unit, responding quickly, efficiently, and effectively to emergency calls within the 40-square-mile district.

For administrative purposes, the city of Chicago is divided into six fire districts of 40 square miles each. A district chief, who is assisted by three deputy district chiefs, heads each district. There are also 12 battalion chiefs who direct the operations of 600 members assigned to 25 fire companies in 3 platoons that work 24-hour shifts.

In June of 1989, I was appointed district chief of the Fourth District. Located on Chicago's West Side, the Fourth District is home to some of the poorest communities in the United States. Whites makeup 85 percent of the department's members assigned to the district. Two percent of these members are female.

As the administrator of the district, I successfully implemented department policies as well as the labor contract that established areas of responsibility for the various ranks. I spent a lot of my time resolving interpersonal conflicts. As I worked to unite employees and maintain a highly motivated, cohesive workforce, I was guided by the concepts and principles described in this book.

These concepts were originally developed for the business community but can be adapted for use in any environment. The fire service, for example, has a semi-military culture, so I made specific modifications for that type of environment. I encourage you to make the appropriate adjustments to fit your own culture. The general concepts and guiding principles are applicable in all types of cultures and serve very well as a starting point to resolve relationship problems in any environment.

Some of the things I have read by race theorists on both sides of the race issue seem to be highly emotional, violent, self-defeating, or hopeless for one side or the other. Of course, I don't have all the answers regarding the complex issues of race and gender conflict. I do, however, suggest that the reader at least consider the concepts presented here as a beginning point toward finding a solution that will serve the best interests of all.

This book describes some of the problems many of us have relating effectively to other human beings. The story of the "Ten 7-Ton Elephants Standing in a Row" illustrates how past experiences often retard our progress rather than provide a basis for a progressive lifestyle. The concept of creating a personal agenda offers a way for each of us to protect ourselves from compromising our values as we attempt to fulfill the objectives of our personal agendas. In discussing the human endowments, I describe the gifts we all have at our disposal to improve the quality of our lives and the lives of our loved ones. There is also a discussion and some sound advice about resolving conflicts. The training models on human behavior, communications, and behavior/attitude/value are presented in clear, plain terms for nonprofessionals and include steps for motivation designed to increase the reader's effectiveness when working with others.

I have also included a chapter about the Million-Man March, which illustrates how the concepts and principles presented here can be applied on a large scale with overwhelmingly successful results.

The book concludes with a summary that reinforces the concepts and principles regarding the connection between responsible behavior and overall well-being. It underscores the fact that our humanity is our common ground and is, in a very real sense, the key to success and happiness.

The no-fault resolutions to racial and gender problems will set this book apart from other texts by identifying freedom of choice, imagination, self-awareness, and independent will as qualities basic to all human beings. The nature of these endowments implies that each human being is accountable and responsible for the effective use of those qualities that make them human and the behaviors that result. It logically follows that these endowments also entitle each person to the respect and dignity due them from their fellow human beings. Rewards and punishment therefore should be based on behavior, not race or gender. One human being does not have the right to judge the innate value of another human being.

Real-life leadership and diversity issues are addressed in the text. These issues come from participants who have attended the author's workshops. In-depth discussions of these concerns are described in the text.

Today, more so than any other time in our history,
we must rededicate and restructure our lives
without minimizing the sacrifices made
by our friends and loved ones.

Chapter One

LET THE HEALING BEGIN

New York City: Dr. Williams

ON SEPTEMBER 11, 2001, the United States of America, specifically New York City and Washington D.C.—the money and the might—were viciously attacked. On that morning, what no American thought possible occurred on American soil. Planes were hijacked and flown into occupied office buildings by individuals focused on the destruction of property and human life. No demands were ever made, and no sense of justice can ever repair the feeling of loss that has been inflicted upon the American people.

New York's two World Trade Center towers and the Pentagon were the specific targets that day. A fourth plane, as we all know by now, crashed in Pennsylvania after passengers apparently revolted against their kidnappers. Ironically, on that beautiful sunny day—where, in New York, the temperature peaked at 82 degrees—there was not a cloud in the sky. By all accounts, it was a gorgeous day. It was a day designed for picnics and pleasantries. What happened that morning was the last thing anyone would have expected.

As word spread regarding the destruction, rescue workers rushed to the scenes. After the carnage began, the actions of these brave men and women would inspire a nation. In New York, where the focus of the world would eventually shift—live and in technicolor—Tower I collapsed. Minutes later, Tower II collapsed, broadcast on every television station in the world. As each tower fell, people watched with a combination of fear and bewilderment that sometimes verged on hysteria. The only thing that could make the tragedy worse, the only thing that would make the sense of loss almost impossible to bear, was the fact that when the towers buckled, those rescue workers—representatives of police and fire departments from all over the city—were inside the buildings as well. And in the wake of the mayhem, those civil servants' actions would become a topic that almost everyone would talk about. While the final body count is estimated at more than 3,000, many people especially sympathized with and prayed for the fire service. Why did so much of the focus shift and remain on those heroic people as opposed to everyone else killed that day?

The answer is not very complicated. In a world gone mad, where violent crimes and mayhem have proven they can touch us at anytime, anyplace, we look for heroes. The warriors in the fire service are the ones who face-off with the dragon, knowing that, live or die, the monster must be extinguished. On September 11, New York's finest and New York's bravest ran into the towers with one mission in mind: respond, and rescue the innocent. As the sirens roared through lower Manhattan, one can imagine the sense of comfort that must have bubbled in the stomachs of those trapped by debris and surrounded by flames. The sense of fear that permeated the airwaves was a feeling that the rest of the world could see, hear, and taste. As the towers fell, killing thousands, the world felt the pain and despair of New York's people. Of course, nothing was more devastating to the spirit of the people than the reality that the heroes—those sworn to assist—could do nothing more than become victims themselves.

Tragedy has many forms. The question is, "How do we, as human beings, cope when the spirit of the nation is cracked and reeling, and its people are afraid and unsure of the future?" While Washington and the rest of the country have been obviously affected by September 11, no place has been more touched by destruction than New York City. While feelings of dread and despair loom across the nation, New York could easily have been left with an even more perverse feeling—defeat. For the average New Yorker, a sense of defeat would be an alien sensation. Like the familiar song, many New Yorkers believe that, "If you can make it there, you'll make it anywhere." In the city that never sleeps, empty streets, barren sidewalks, abandoned shops, and, worst of all, silence (except for the never-ending squealing of emergency vehicles) speaks to the horror that tightly grips the chest of every single New Yorker's heart.

The city reeks from the smoldering ash and fires that as late as December 2001 were finally burning themselves out. People are still without housing in lower Manhattan, and businesses have closed at a record pace. But life goes on and people fortunate enough to still have jobs return to the sidewalks and subways. There they are confronted by the pictures, usually happy poses of the "missing", plastered on every light pole, doorway, and subway station. The shared pain of almost 9 million residents and their neighboring tri-state brethren who come there to work, is as discernably thick as the mushroom cloud that hung over the city until mid-October.

As the rest of the world shows its support for New York by sending cards, money, food, and clothing, New Yorkers valiantly march forward. Yet, this place, America's most densely populated, most powerful, and most influential city, no longer stomps with its usual bravado. Some people are wary, watching every plane that soars through the sky. City officials question the rationale of rebuilding the towers, possibly thinking that, if redone, they should be "maybe not so tall this time". This is a sentiment shared in other cities, such as Chicago, where people are fearful for their own architectural wonders.

Chicago: Chief Crane

On that morning, I walked into the training room of the Elk Grove Fire Department, a suburban community in Illinois northwest of Chicago. I was prepared to conduct a leadership class for the Mutual Aid Box Alarm System, Division 1 Fire Chiefs. As I began to set up, I saw people standing in front of a television set. One of them, a photographer named Roy Hervas, came up to me and introduced himself. Roy had agreed to take pictures and video of my presentation that day. He asked if we should change the way the chairs and tables were arranged.

After some thought, we decided on an arrangement that suited both of our needs. As we finished the setup, he asked me if I had heard what was going on in New York. "No, what happened?" I asked. He told me about the airplanes that had flown into the World Trade Center. It was at that point that I remembered seeing a plane flying into a building on television as I entered the room that morning. I thought someone was playing a video game. Simply, he told me "No, that was no game." Incredulous, I asked him, "Are you kidding me?" He assured me he was serious and that this "was no kidding matter". That got my attention.

As I watched the firefighters on television walk down the street toward the towers with civilians hurrying in the opposite direction, I reflected on the countless times I had prepared to enter fire buildings. Those in the service seem to agree that a certain feeling builds as you prepare for battle; your entire body goes into a state of high alert. I would compare it to an animal-like survival instinct that wells up and dominates the senses. This instinct is not so much for one's own personal protection, but for the survival of any living thing that may be in danger. Stories of firefighters saving cats and dogs are legendary. These tales are significant because all life is worthy of risk to firefighters—all life.

As I continued to watch the television, I observed the firefighters entering a decimated high-rise building filled with people. I knew instantly that the major challenge they faced was life hazard. There is nothing more important than life safety in the fire service. Time and energy are critical. Those firefighters could not afford any wasted motion. Put simply, there must be a positive return for every action taken. At a time like September 11, training and past experiences come heavily into play. Basically, these are the key elements that separate firefighters from the citizens they serve. Because of this prevailing sense of safety and rescue that firefighters possess, I have continually asserted for more than 40 years, that the firefighter is the clearest personification of what we call the "Noble Breed".

What we know for sure is that each one of those firefighters felt that their presence in the towers would make a difference. If that were not true, those firefighters would have been walking in the same direction as the citizens. Watching the horror unfold on television, I did a quick calculation and estimated that there would be about 30,000 people in those buildings. If that were the case, it would not be unreasonable to expect that there would have been approximately 10,000 deaths as a result of the collapse. It was nothing less than a miracle that fewer than half that number died. Undeniably, the quick and effective work of the emergency workers played a role in the resolution of the catastrophe. The news accounts claimed that over 25,000 people were able to escape on that day and that around 341 firefighters lost their lives. There is no doubt in my mind, or in the minds of those who watched, that the "Noble Breed" made a tremendous difference in the outcome of that day. Yet, the loss of my comrades is extremely painful. My life has not been the same since. A reality that I am sure speaks for many people around the world.

Immediately, cell phones and pagers began to go off as communities reached out to locate the high-ranking emergency service officials surrounding me. Obviously, the scheduled class could still be taught, but the uncertainty was overwhelming. As word filtered in about the Pentagon, as well as the plane in Pennsylvania, rumors spread that Chicago's Sears tower—the world's second tallest building—was the next target.

As plans to evacuate the tower were set in motion, I made my departure and headed back into the city. Feeling on edge, I decided to stop and visit with some friends and continue to watch the news. We all agreed that there was no way to control a person who was prepared to die and take others with them, except to possibly convince them that there had to be something good that they could try at least one time before leaving this world. I later called Michael Buren of Northwestern University to see if the class scheduled for September 12, 2001 was postponed. He informed me that I was expected to teach as planned.

When I arrived on the 12th, I learned that the students I was meeting had had another class cancelled the previous day. Because some of the students had traveled in from out of town and had made hotel commitments for the week, it was important that the classes go on if at all possible. I immediately noted that the students there were going through the same type of listlessness that I found myself experiencing. I did not believe I was ready to teach this four-hour motivation workshop for fire service supervisors. After confessing to the class that I needed more personal motivation than I thought I could provide, we agreed to lean on one another and work our way through the session. We began by holding hands in silent prayer. As the clock ticked onward, there was a sense of unity that flowed from each of us. This unity created a bond that we all fed on. There was a sobering seriousness of purpose that lingered throughout the room. We were all hurting, and our needs were great.

At the end of the class, I thanked the students for their support and complimented them for their diligent and heartfelt participation. Later that evening, Mike called to ask if I would be able to meet with the same group the next day to conduct a leadership workshop. The scheduled instructor was unable to attend because he had been called to an emergency meeting on weapons of mass destruction. I agreed to teach the class. After the interaction of the previous day, it was warming to see the same group of people again. We were all a little better as a result of the previous day's experience. It was as if we belonged together.

Yet, as the week went on, I continued to be listless. It was as if I was coming down with the flu. I had little energy, just barely enough to do the bare necessities. Later that week I learned that Chief Robert Hoff, the director of training for the Chicago Fire Department, and Lieutenant Richard Kolomay of the Schaumburg Fire Department, along with 75 other firefighters and support personnel had gone to New York to assist in the rescue effort. Chief Hoff and Lt. Kolomay are authorities on special rescue procedures. They had extensive experience working with friends who are members of the New York Fire Department.

In the meetings we had following their return, it was evident that the death and destruction they had witnessed in New York had a profound impact on them. Roy Hervas, the photographer, said that there were pictures he did not take because they were too gruesome. Kolomay and Hoff said it was important for them to have someone they could share their stories with who could empathize with them as devoted rescuers. Both of them expressed concern for the other rescuers who could benefit from a solid support system upon their return home. We all felt that only time would heal us, and, as time passed on without another incident, we would feel better and get on with our lives. I don't know what the new "normal" will be, but I do know that the old normal is gone forever. This truth holds true for everyone, especially those in the fire service.

It was Lt. Kolomay who encouraged me to continue writing about the importance of establishing and maintaining solid relationships with others. Since September 11, relationships with family and dear friends have become more priceless than ever. These relationships, plus decisions to alter our lives in this new world, appear to be the key to moving forward and reclaiming our sense of well-being. I was only able to shake my own feelings of lethargy by redefining my personal agenda, which we will outline in the following chapter.

New York: Dr. Williams

Terrorism is terrifying; there is no denying this. Bad things happen to good people; this too is a tragic fact. And, unavoidably, everyone dies. Some things in our lives are out of our hands.

For the most part, people walk through life hoping for the best. But, and this is no small "but", what we can do is determine how we will behave in life, as opposed to waiting for disaster. In order to overcome our fears, we must understand something: anticipating tragedy is a complete relinquishing of real living.

Waiting for the next wave of terror to arrive will cause fear to overwhelm our every sleeping moment, our every waking moment. Tragically, allowing oneself to succumb to a primal sensation like fear—which eventually transforms itself into an even more debilitating depression—is something most people feel they cannot control. The east coast continues to wonder how, if ever, they will be able to function normally again.

The fight that every American must engage in is not with terrorism. As Edward W. Said asserted, "President Bush can no more rid the world of evil-doers than he can stock it with saints." The job of repairing our wounds cannot come from money or military might. The presence of the Army and National Guard—those soldiers stationed in the subways and bus stations attempting to walk among the citizens normally armed with M-16 assault rifles—can provide only a limited sense of physical security. The refurbishing of our spirits must come from our understanding that how we live—no matter how long or short our time may be—must encompass a knowledge that we alone are responsible for our own happiness and well-being. Think about it. Isn't it the job of every parent to cultivate the human endowments of little people so that they can always strive to be winners? Don't all parents have a responsibility to let their children know that their personal desires and objectives are attainable if they choose to pursue them? Obviously, forces will always attempt to make us believe that we do not control our own lives. But, a bully is only as strong as we allow them to be.

All maturing minds have to do, all any of us have to do at any time, is to choose who we are going to be and what that means to us and those we value most. What we cannot control, we cannot spend our energy dwelling upon. As we now see, life is too short to waste time on non-effectual matters.

Crying over death is natural; crying over life is a choice. Destroying buildings and killing innocent people are more than just horrific acts. If we agree that the purpose of terrorism is to terrify, then we must admit that there can be no more obvious attempt to usurp our sense of responsibility for our own behavior than what happened the day those planes were turned into weapons of mass destruction.

Living in fear—a state of continual agitation and panic—leaves us stagnant and immobilized. If we succumb to the conflict, we relinquish our ability to use our imaginations to see beyond the present turmoil and live according to our own interests. Surely the war we wage within ourselves, as well as with an unidentifiable enemy, will never bring back those so abruptly yanked from our lives. Our principles and objectives should guide us into the future.

The agenda of terror belongs to others; let them keep it. For those of us left feeling devastated by the loss of family, friends, and strangers, as well as those who just find themselves plain old afraid of what might or might not come, a personal agenda is definitely needed—now more than ever. Many of us have been forced to adjust our agendas after September 11, but, thankfully, we can also choose to do that whenever we deem it necessary. For those of us who have never considered developing an understanding of how we live, we must take hold of our highest values—that perfect painting of ourselves—and seek to live accordingly.

The most important human endeavor is the striving for morality in our actions. Our inner balance and even our very existence depend on it. Only in our actions can [we attain] beauty and dignity in our lives.

~Albert Einstein~

Racial and ethnic profiling has also taken on a new face since the attacks in September. In the midst of our fear, we Americans must continue to hold ourselves accountable for our words and actions—recognizing that looking and dressing a certain way does not make one the enemy. This sentiment rings with a special tenor of importance in New York, where so many Middle Easterners live and work throughout the city.

September 11 was one of the most extreme examples of an inhumane act. We may choose to cope with the events of that day by relinquishing the gifts of our humanity. However, when we turn away from these gifts, we dishonor our gifts, our fallen, and our heroes. We must instead turn to those nearest and dearest while we follow our own path towards a newly defined self-fulfillment.

Chicago: Chief Crane

I am often reminded lately of Raymond E. Orozco, who as the Fire Commissioner of Chicago, was adamant about including family members in department ceremonies. Family members would be asked to pin the badges on newly promoted members while department photographers took their pictures. Commissioner Orozco would tell the audience that the fire department was their extended family. What is so compelling about this, in light of "9-1-1", is that this inclusion of the families allowed them to participate intimately in the world of their loved ones who have dedicated their lives to strangers. This sense of giving among firefighters is something that the average citizen will never witness or experience firsthand.

The incidents of September 11, 2001 drove that fact home in a very personal way. People from all over the world came together to lend support in any and every way they could. It seemed as if each of us was driven deeper into our inner being and put in touch with our highest good. This led us also to communicate with the highest good in others. And, although there were unfortunate attacks on innocent Muslims and Middle Easterners, the vast majority of citizens quickly denounced those heinous acts. The increased level of caring and national pride has been extremely heartening.

It would be a shame to allow this caring and pride to be lost. It is this effective communication with our inner souls that is at the heart of our personal agendas. Keep in mind that the objective of the personal agenda is to put our outward behavior in harmony with our heart-of-hearts. This behavior is the essence of personal empowerment.

New York: Dr. Williams

As the comrades of those fallen warriors continued to search through the tons of rubble with bloodied hands—images that for months the media zoomed in on with disturbing regularity—the question of coping for those left behind seems to be the call-of-the-day. Listening to veteran firefighters discuss the minutes leading up to the collapse of the towers has provided clear evidence of the conflicts these fellow warriors continue to experience regarding the actions taken that morning. Interestingly, one former chief said, "The only sensible command would have been to 'stand-down'." But he immediately countered with: "But that would have been impossible. It goes against all of our training, all [of] our instincts." He continued, "There was a fire; there were victims; no [person] who wears the uniform could have done any less." At this point, this former leader of identically brave firefighters said something that put it all into perspective: "They died doing what we do." As sad and final as that sounds, every person who sacrificed their life on September 11 proved that, no matter what, they had embraced this path as a part of their life's agenda—a part that none of them could have denied.

In the middle of the mass hysteria, we all saw the images in New York of planes exploding, people dangling out of windows over 100 stories high, as well as the running and screaming. But most memorable were those brave individuals in matching uniforms, rushing in to save lives without hesitation. That morning we watched the "Noble Breed" enter the carnage, never intending to leave their own families and friends, but always knowing in their subconscious that there existed the most minuscule possibility that their training and preparation would not be enough, that they might not return home. This reality is, as the chief expressed, because "it's what we do".

Our heroes in New York and nationwide continue to do their jobs. Policemen still patrol and respond. Firemen still rescue those trapped and engage the unmerciful phenomena known as fire. Yet, reports say that many of New York's bravest will retire soon and even more are questioning whether they can go on. For those who feel uncertain, pondering a few simple questions might help. At the end of this chapter, we have included a set of "Questions for Discussion". Those reconsidering their futures should pay particular attention to questions 7 through 14.

In Times Square, a huge bronze statue sits in the street covered with flowers and cards. This monument is constantly visited, frequently touched, and undeniably moving. The statue, which stands about 15' to 20' tall, is of a kneeling fireman, helmet at his side, face in his hands, weeping for his lost comrades. It expresses a sentiment felt by everyone who passes—an awareness of loss that flows through the veins of every dedicated officer in a uniform.

This statue, along with every firehouse in Manhattan is now open to the public. The firehouses have been made into shrines with cards, flower reefs, pictures of the fallen, and an obvious fog of sadness. These memorials help us all remember that tragic day, and in some way, they help heal the pain. And, as every city billboard is plastered with new slogans like: "Everyone Needs Heroes. Thanks New York!" A sense of appreciation and normality is slowly filtering into the mainstream thinking. But, with anthrax scares and a media bombardment of the American "warring abroad", we can no longer simply click the channel to get away from the usual reports of bombings in Jerusalem and Bosnia. What we should notice is that we now see life differently. Not only our own, but the lives of other human beings as well. We now recognize that this type of horror and tragedy exists around the globe and that families everywhere lose loved ones because of violence, and there is nothing that will ever bring them back. With this awareness of other human beings' humanity and courage, we realize that we, like them, will survive.

QUESTIONS FOR DISCUSSION

1. How can the people in New York, and worldwide, ever find peace again?
2. How can we heal—collectively and individually?
3. Can we learn to live with fear? More importantly, should we?
4. And what do we do when our heroes fall, when our "finest" and our "bravest" become victims like those they so valiantly attempted to save?
5. How can we prevent ourselves from being afraid?
6. How can we live a normal life while surrounded by physical ruin and emotional despair?
7. As a member of the fire service, am I as physically prepared to do my job as I can be?
8. Am I always mentally focused on the task in front of me, or are there factors that could distract me and possibly result in me, or those who depend on me, getting hurt?
9. Have I chosen the path best suited for who I am today?
10. Do I know that I am lacking in some area vital to my performance? Am I willing to acknowledge this and seek retraining, or should I remain quiet and hope for the best?
11. Do I believe that those under my command are properly prepared?
12. Do I believe that those in command positions have the well-being of the rest of us in mind?
13. Am I a winner or a loser?
14. What's next?

"Every heroic act is beautiful, and causes the place and the bystanders to shine"

~Ralph Waldo Emerson~

Chapter Two

PERSONAL AGENDA: NOW MORE THAN EVER

In today's world, a personal agenda is the key to overcoming pain and moving forward. There is no more power that exists than in empowering the self. Establishing our values and priorities is of particular importance for those who find themselves feeling defeated after the events of September 11, 2001.

WHY DO WE NEED A PERSONAL AGENDA?

WHETHER A CONTAINER is viewed as half-full or half-empty depends on the intentions, or objectives, of the viewer. Those individuals who have assumed the responsibility of emptying the container may posit that the container is half-empty. By contrast, those individuals who have assumed the responsibility of filling the container may see it as half-full. These different perspectives are based simply on the things that are important to the individual. If you change the objective, you change the perspective. Generally, when people look at the same situation and come away with different interpretations, it is because what they see is based on their own objectives. Usually those with similar agendas will come away with similar interpretations.

While people, as a matter of course, prepare a well thought-out agenda for running a professional or organizational meeting, many individuals do not typically put the same kind of careful thought into developing an agenda for living their personal lives. As a result of their failure to create a personal agenda these people will eventually and inevitably assume either an agenda assigned by another individual or one sanctioned by society. Although people may haphazardly meet some of their own objectives during the process of following someone else's agenda, complete satisfaction and attainment of their goals will elude them. This is because goals must first be identified and understood before they can be strived for and reached. People cannot achieve happiness and success without first defining what happiness and success mean to them.

Designing your personal agenda

Your personal agenda should identify the principles that will guide your behavior as well as define what role you will play in your relationships with the people in your life who are most important to you. Then, taking this information into account, you can begin to set your objectives and develop a plan of action. Establishing your goals and planning your course of action enables you to have a clearly defined personal agenda. This agenda keeps your external behavior in accord with your internal objectives—creating a sense of internal harmony, as opposed to the disjointed and somewhat powerless feeling of moving through life reacting to someone else's program. Eventually, as a result of implementing your own personal agenda, you will become proactive, seeking out opportunities to fulfill your stated goals in life. If you are still a little confused, do not panic. Try this exercise:

1. Sit down with a pen and some paper. Use your imagination to create a scene five to ten years into the future. If that scenario seems impractical, set the scene for next week or next month. Plan to invite five of the most important people in your life over for a meal. Make this gathering a testimonial dinner for you.

2. Next, write down your guests' names and their relationship to you. Write out their testimonies to you; in other words, write down the things that you would like each one of them to say about you. Since there will be five guests, there should be five separate statements. You may invite your missing loved one, too, as a special guest and write a speech for him or her.

3. From this point, identify the principles that you would like to guide your behavior and write them down. These should be timeless and changeless concepts that are true and important. Together these principles and testimonial statements form the core of your personal agenda.

4. Put the statements and principles into a narrative form. The narrative should be a fluid expression of the ideas and goals most important to you. Your personal agenda is based on the contents of these written statements.

Now, go forth and live your life in such a way that you will be worthy of the testaments given at your testimonial dinner. Remember that we are all endowed with inherent gifts that help us simplify what may at first seem complex, gifts that empower us to work toward fulfilling our goals.

Chapter Three

THE HUMAN ENDOWMENTS

THE HUMAN ENDOWMENTS of self-awareness, imagination, independent will, and freedom of choice have been identified as gifts that come with birth. And, because it is an essential part of the internal harmony that comes with personal empowerment, a person's conscience, though not present at birth, is also considered a human endowment. We cannot have internal harmony without a clear conscience. Conscience is a learned awareness of morality and ethics developed and learned over time from authority figures.

Defined separately below are definitions for the endowments. The endowments are tools that all of us have and can use to meet our personal objectives and to protect our interests.

- **Self-awareness** is one's ability to evaluate his or her circumstances and to determine what changes are needed, if any.
- **Imagination** is the ability to create images in one's mind beyond his or her present reality.

- **Independent will** is the ability to act based on one's self-awareness, independent of all other influences.
- **Freedom of choice** is the freedom to choose one's response to any event or stimulus that comes from the outside, including the freedom to choose not to respond to the event or stimulus.
- **Conscience** is a deep inner awareness of right and wrong. It is a sense of the harmony between our thoughts and our actions and the principles that govern our behavior.

PUTTING YOUR GIFTS TO USE

Now try to think of ways that you can use these inherent human endowments to help bring about effective, positive changes in your life. Start by reviewing your personal agenda. Use your gift of self-awareness to identify the things on your agenda that you can act on right away, and then follow through by taking immediate action. Use your gift of imagination to help inspire you to meet the objectives on your agenda that require more preparation, resourcefulness, or creativity. Use your gift of independent will to make decisions based on whether or not they can help you move toward your stated goals, as well as to aggressively look for opportunities in the larger community to turn your aspirations into achievements, your dreams into reality. In situations where you're compelled to react to someone else's behavior and actions, use your gift of freedom of choice to respond in accordance with your own principles and objectives, not theirs. Do not allow yourself to slip into old habits of behavior based on following or reacting to someone else's agenda. And finally, use your gift of conscience to guide your decisions and actions to ensure the most positive results for yourself without bringing harm or negative consequences to others.

I do not intend to imply that your endowments should be used in any particular order or applied uniformly in all types of scenarios; essentially, they are yours to use as you see fit. Be thankful for these gifts, and do use them—they are your keys to self-empowerment.

FINDING THE POSITIVE AND TAKING CHARGE OF YOUR HAPPINESS

When you feel empowered, there is no limit to what obstacles you can overcome and what levels of happiness and success you can attain. Prisoners of war have related inspiring stories of how they used imagination and freedom of choice to help reduce the effects of the terrible treatment they received at the hands of their captors. While the movements and actions of a prisoner are restricted, the gift of imagination is not subject to outside interference. So, prisoners have two choices: they can bemoan their imprisonment, or they can actually take advantage of it by using the time to prepare for the future. One man, for example, used his gift of imagination during his captivity to design his dream house. After his liberation, he returned to his hometown and built the dream house, creating something positive from an otherwise negative experience. Similarly, enslaved Blacks working in the fields sang spirituals to help ease the pain of the harsh reality in which they lived.

So, it is vital to understand that others may restrict your movements and your liberties, but no one can control your imagination or your freedom of choice. Take a look at the next example and think about how it further illustrates the strength that each of us has within us to confidently direct our lives and respond to whatever dilemmas we are faced with.

An angry band of White men came to the home of a Black family in the segregated South of the 1920s. The group of men demanded that the father send one of his sons outside to be punished because he had been seen running amuck and willfully destroying part of a watermelon patch. The father told the leader of the group that he wholeheartedly agreed that such behavior should not be tolerated and that he intended to punish his son for his actions. He assured the men that he would see to it that his son understood the seriousness of what he had done and that nothing like that would happen again. The men took the father at his word and went away.

This incident illustrates how self-empowerment (in this case, the father's) can turn an extremely volatile situation into a reasonably straightforward interaction in which conflict is resolved. The father's clear-headed and self-assured handling of the situation compelled the angry group of White men to listen seriously to what he had to say, and his willingness to take responsibility for his son's actions defused their anger. The internal harmony that comes with personal empowerment allows us to circumvent, as well as rise above, racial and gender concerns. Self-empowerment is the ability to choose and manage change; it gives us the courage and the energy to redirect and self-direct our lives.

THE PERSONAL AGENDA: GUIDING PRINCIPLES

Our responsibilities for and to ourselves include the following:

- Maintaining our physical and mental well-being
- Identifying and fulfilling the desires of our heart-of-hearts in order to achieve personal happiness
- Behaving in accord with our highest-held values and vision of ourselves
- Being accountable for our behavior at all times
- Using our human endowments as tools to meet our responsibilities and goals

NOTE: *While using this guide, keep my rule of engagement in mind—protect your interests at all times without doing any unnecessary harm to others.*

Personally, I perform specific activities to ensure my success in meeting these responsibilities. They include diet, meditation, prayer, and exercise.

PERSONAL AGENDA: SUMMARY

It is important that you develop your agenda from your personal values because this enables your outside behavior to be in accordance with your internal desires. Some of us feel that this balance contributes to our overall personal comfort and well-being. When we behave in concert with our innermost beliefs, we tend to be at ease with ourselves. We are happy when we do what we really want to do and it turns out well. It is a self-fulfilling experience. Conversely, the absence of such harmony between our outside behavior and our internal beliefs results in our being uncomfortable with ourselves. Many people are unhappy even at the sight of themselves in the mirror, experiencing depression and becoming despondent as a result of the conflict between their actions and their beliefs.

Your agenda should serve as a guide for your behavior in both your personal and professional lives. You are now to seek out opportunities to take action and behave in ways that will make those statements true. Each one of us has the ability and the responsibility to control our behavior. When the people who are important to us say good things about us, we feel good about ourselves. You may plan a testimonial dinner for yourself as often as you like and you have complete editorial rights over your personal agenda. You can change or adjust it as you see fit. Life is in a state of constant change, so it would follow that your agenda should change also. Whatever the circumstances of your life or the stage that you are at, your agenda should always be a source of motivation for you.

As a firm believer in the motivational power of personal agendas, I would like to share with you my own personal agenda, which I developed using the process detailed in the worksheet (Appendix A):

Be supportive of others. Convey to others that life is worth living and sharing, and believe that there are answers to most problems. Become all that I can be, and help others become all that they can be. Try to serve, or at least recognize, the highest good in others and myself. Worship God by serving others in his honor. Maintain my health with a sensible exercise program and diet.

As simple as it may sound to others, these statements are very motivating to me. When I lose sight of what is important to me, it is good to have my agenda to use as a touchstone. The five people that I identified as being most important in my life are my mother, wife, and three children. The statements that I wrote for each of them and the statement of my principles, were the testimonies that I used to create my agenda. It is interesting to note, by the way, that my agenda serves me equally well in relationships I have with people outside of this circle of five. The sense of well-being and increased comfort with myself transfers into a more relaxed level of participation in my relationships with others.

REVIEW OF THE PROCESS

1. Plan your testimonial dinner to be held next week, next month, or five or ten years from now.
2. Identify the guests you will invite and the role each one plays in your life.
3. Write out what you would like each guest to say about you.
4. Write out the principles that you want to use to guide your behavior.
5. Combine the statements and principles into narrative form and you will have your own personal agenda.

Now, using the principles you have chosen to guide your life, seek out opportunities to make the content of those testimonial speeches true.

THE TEST

Periodically, read your personal agenda and make adjustments to it whenever the circumstances of your life call for it. When you read your agenda, do you feel inspired to continue moving on and to look for opportunities to meet your objectives? If not, try again. Create a new personal agenda by repeating the steps of the process; be sure to follow the instructions carefully and to be completely honest and thoughtful in your responses. If you do feel inspired after reading your agenda, congratulations! You can use this guide throughout life as your chief resource for finding the motivation, strength, and confidence to pursue your goals and to achieve personal success and happiness.

STRESS MANAGEMENT

Because there are not many vocations more stress producing than the fire service, we have included this section on stress management. For your convenience, here are two stress-management models that you may find useful. I use the 2-2-5 technique regularly to relax.

Meditation model

Ignore the distracting thoughts or ideas that you think of while doing this exercise. You may consider them after you have completed the exercise. Try doing this exercise twice a day for 20 minutes to minimize the effects of stress.

1. Choose a number from one to nine.
2. Sit in a comfortable position with feet flat on the floor, arms resting on your lap.
3. Close your eyes, keeping your mouth slightly open to keep your teeth from touching.
4. Repeat your number over and over to yourself.

2-2-5 Relaxing technique

This technique should be performed while sitting with both feet flat on the floor and both hands resting in your lap. It is important that you feel comfortable.

1. Visualize and repeat the following:

My body is relaxing, completely relaxing,
and I am going into a deep, deep sleep—into a deep,
relaxing sleep. I am in a deep, relaxing, refreshing sleep.
I am going to do my 2-2-5 and I am going to sleep
for three minutes. While I am asleep, I want my subconscious
to remove all stress and fatigue from my body.

(Other suggestions may be inserted at this point)

Each and every time that I suggest sleep to me, I will sleep
quicker and deeper than the time before.

2. Close your eyes and breathe through your mouth. Your mouth should be slightly open throughout the breathing technique.
3. Inhale, hold it, and count: "1, and... 2, and..." Exhale completely.
4. Repeat.
5. Inhale, hold it, and count backward from: "5, and... 4, and... 3, and... 2, and... 1" Now, exhale completely.
6. Inhale and exhale normally. When you are ready to awaken, repeat the following:

As I prepare to awaken, I will count from one to three.
On three, I will open my eyes slowly and awaken.
I will be in a new state of normalcy,
completely free of stress and fatigue.

7. Count: "1, and... 2. On 3, I will open my eyes and awaken. 3." Open your eyes and awaken.

2-2-5 Learning technique "A"

This technique should be performed while sitting with both feet flat on the floor and both hands resting in your lap. It is important that you feel comfortable.

1. Repeat the following:

> *My body is relaxing, completely relaxing, and I am going deep, deep asleep. I am going to do my 2-2-5 and I am going to sleep for three minutes. While I am asleep, I want my subconscious to remove all stress and fatigue from my body.*

2. Visualize the following:

> *The material that I am about to study will become deeply impressed upon me and be permanently retained by me. Each and every time that I suggest sleep to me, I will sleep quicker and deeper than the time before. At the end of three minutes, I will awaken feeling more relaxed and more refreshed—more relaxed and more refreshed than ever before. My body will be in a new state of normalcy, completely free of stress and fatigue.*

3. Close your eyes and breathe through your mouth. Your mouth should be slightly open throughout the breathing technique.
4. Inhale, hold it, and count: "1, and... 2, and..." Exhale completely.
5. Repeat.
6. Inhale, hold it, and count backward from: "5, and... 4, and... 3, and... 2, and... 1" Exhale completely.
7. Inhale and exhale normally. When you are ready to awaken, repeat the following:

As I prepare to awaken, I will count from one to three. On three, I will open my eyes and awaken. I will be in a new state of normalcy—completely free of stress and fatigue.

8. Count: "1, and... 2. On 3, I will open my eyes and awaken. 3." Open your eyes and awaken.

2-2-5 Learning technique "B"

This technique should be performed while sitting with both feet flat on the floor and both hands resting in your lap. It is important that you feel comfortable.

1. Repeat the following:

My body is relaxing, completely relaxing, and I am going deep, deep asleep. I am going to do my 2-2-5 and I am going to sleep for three minutes. While I am asleep, I want my subconscious to remove all stress and fatigue from my body.

2. Visualize the following:

The material that I have just learned is now permanently retained and will be easily recalled when I need it. Each and every time that I suggest sleep to me, I will sleep quicker and deeper than the time before. At the end of three minutes, I will awaken feeling more relaxed and more refreshed—more relaxed and more refreshed than ever before. My body will be in a new state of normalcy—completely free of stress and fatigue.

3. Close your eyes and breathe through your mouth. Your mouth should be slightly open throughout the breathing technique.
4. Inhale, hold it, and count: "1, and... 2, and..." Exhale completely.

5. Repeat.
6. Inhale, hold it, and count backward from: "5, and... 4, and... 3, and... 2, and... 1" Exhale completely.
7. Inhale and exhale normally. When you are ready to awaken, repeat the following:

> *As I prepare to awaken, I will count from one to three. On three, I will open my eyes and awaken. I will be in new state of normalcy— completely free of stress and fatigue.*

8. Count: "1, and... 2. On 3, I will open my eyes and awaken. 3." Open your eyes and awaken.

Chapter Four

EMPOWERMENT ~VS~ ENTRAPMENT

THE LAST REPORT THAT I RECEIVED stated that more than 80 percent of the firefighters in the United States were volunteers. The challenges facing the leaders of volunteer fire and emergency services are similar to those that face the religious leaders in our communities. Leaders must attract and maintain memberships that will contribute their resources and time to serve their fellow citizens without compensation. The volunteer firefighters are expected to provide emergency services in the worst of conditions and circumstances. The simple truth is that fire burns with the same intensity in a township staffed by volunteer firefighters as it does in the Bronx in New York or the west side of Chicago. Hostile fire is the enemy wherever it occurs. The enemy does not make any distinction between volunteer or paid firefighters, and it is foolish for us in the business to make such a distinction.

The concepts discussed in this book will work equally as well in the small communities served by volunteer firefighters as they do in the larger cities served by paid firefighters. In many instances, these concepts serve the leaders of smaller communities better because they provide solutions to interpersonal problems based on proven human behavior skills without regard for compensation. It matters little if the firefighters are paid or volunteer—they are all human beings. As human beings, they bring a set of unique gifts and needs to our organizations. Organizations exist to utilize the gifts and serve the needs of their members.

It is my belief that the American worker is the most productive worker in the world. However, insecure managers and supervisors often fail to prepare their task providers for success. Supervisors must take the time to inform workers of what is expected of them and give them the training and support necessary for their success. When this happens, production and morale are maximized. The employee comes out of such an experience with a new sense of commitment to the organization. The failure of managers to provide this type of training and support results in frustration and minimal production from employees. While there has been improvement in this area, both the public and private sectors of American business communities still have more ground to cover in this effort. The major cause of this problem may be the insecurity of supervisors and managers. A supervisor's position may be that new workers "get their skills and knowledge like I got mine". I have heard some express concern that if they trained the task providers too well, the workers would take their jobs. Supervisors with this attitude do not realize that to teach is to learn twice. The supervisor should be the most secure member of his/her work group. Otherwise their insecurities will affect the manner in which they do their job.

Consider this exchange. While I was a lieutenant on a hook and ladder company with the Chicago Fire Department, a relief officer said to me, "Now I know why you train all the time. You are preparing yourself for the next captain's examination." I also had a stated commitment to do as much as possible to get as many of the members who worked for me promoted. The success of subordinates should be a matter of pride for leaders, much like the pride that a teacher may experience with the success of his/her students.

The organization has an obligation to train its leaders first; the leaders then have an obligation to train their subordinates. The fire service has a longstanding practice of promoting people to leadership positions without appropriate training. Sometimes the newly promoted officers are given a white shirt and the sign of the cross and told to move forward to direct and control the members of their units. The white shirt is a symbol of authority in the fire service, while the sign of the cross is a silent prayer for his/her success in their new role. No one would deny the importance of these two items. However, effective training would enhance both of them. The training would give added value to the entire process and prove priceless as the lessons learned were passed on to the task providers. Adding a training element to the promotional procedure is a proactive step toward empowering the workers. Failure to have a training element can entrap the worker and leave the organization in a position of having to react to his/her failures and rescue them from their undirected misdeeds. Simply stated, the organization will be proactive and train their people for success, or be reactive to save them from the results of their failures.

We will either empower our people with the tools needed for success or entrap them due to the lack of those tools. The informal polls taken at my workshops indicate that workers feel comfortable and committed to managers who demonstrate an interest in their success. These polls also show that the majority of workshop participants are dissatisfied with their current supervisors. The polls have been taken for the last 10 years and the number of workers experiencing effective training and positive encouragement has been increasing. Achieving organizational and personal excellence go together. The objectives that you identify personally and the objectives that the organization identifies need not be at odds with one another, but in fact, can be parallel. Many of the skills and lessons learned in the workplace are helpful to us in our private lives.

To commission, authorize, and train is what we mean when we talk about "to empower". "To entrap", on the other hand, is to catch in a trap or a snare, to bring into difficulty, to deceive, or to trick. If we don't have clearly written mission statements, organization policies, practices, procedures, job descriptions, and training, we are preparing our people for failure. Often when this happens, we also play a little game called "gotcha", in which we catch people doing something wrong and hold them accountable for their errors. Errors that often could have been avoided with effective training. The result is that we're always reacting to crises, and always a step behind them. Many times any discipline that follows includes some sort of training.

I recommend that we set up a prototype for being proactive. The most proactive thing we can do in an organization is to train our people and do what is necessary to protect them from failure. The most important responsibility of managers is ensuring the training of their task providers. Responsibility without instruction is an imposition. Holding one accountable and responsible for functions that they have not been effectively trained for is unfair. After effective training, supervisors should seize every opportunity to catch workers doing something right and complement them for a job well done. Certainly no effective manager will ever deliberately bring a worker into difficulty or attempt to trick them. If we operate in a way that does not inform our people of what's important, or what the agenda is, and if we don't properly train them, then we have tricked them. The impact is the same as it would be if we were withholding information.

One of the problems that we had in Chicago several years ago was the withholding of information that would have been useful for test takers seeking promotions. I'm not going to say that this was an effort to trick or deceive, but it certainly had that effect. It wasn't until the federal courts got involved as a part of the affirmative action process that a reading list identifying reference materials was made available to members preparing to take promotional examinations. One of the things that I did when I became assistant director of training was to distribute that information throughout the department. This created a level playing field for all of the candidates up for promotion and contributed to the personal empowerment of all those involved.

Clearly written mission statements or job descriptions that describe the department's commitment to the community and the primary responsibility of its members are tools that empower workers. These types of documents can clear up a lot of issues and prevent confusion within an organization. An example would be for members to understand that the primary objective of the fire service is fire "prevention". Therefore, the advent of a fire indicates that the fire service has failed in its primary responsibility. Firefighters, being the true heroes that they are, walk around in the firehouses chanting "Speaker-speaker on the wall, please give us another call." They anxiously await the opportunity to serve and protect, the opportunity to demonstrate skills that set them apart from the rest of the community. These task providers deserve all the support that their organizations can provide them. The best support that could be provided to them may be an effective fire prevention program.

While firefighting is important and spectacular, it is not our only duty. Mission statements and job descriptions made available to all members will inform them as to what their other responsibilities are and put them into proper balance. Additional functions may include fire prevention inspections, which have proven to be an effective means of not only preventing fires, but also in avoiding untold numbers of injuries and deaths. The fact is that the perfect fire service is one that has an effective fire prevention program and no fires. Increasingly, emergency medical services make up the larger part of the services that firefighters provide.

Supervisors should use department mission statements, policies, practices and procedures, and, where applicable, labor contracts as administration tools. It is important that they are available to all of the members. Managers who are available will likely have fewer complaints about task providers having bad attitudes or being uncooperative. The member who is uncooperative may behave that way because he's not familiar with, or doesn't know, what you're trying to do. Managers who have failed to inform or train employees as to what is acceptable behavior have lost many arbitration rulings. When we communicate solely by word of mouth, some people get information and other people do not. Considering the complexities of verbal communication, it is a wonder that it doesn't cause more confusion than it does. What a speaker says and how he says it depends on his background. What the listener hears and how he interprets what he's heard depends on his background too. It is obvious that people from different backgrounds could have conflicts because of these differences. In my neighborhood as a youngster, we would call a nice automobile a "bad short". In other neighborhoods, a nice automobile would be referred to as a "cherry ride". When these kinds of differences exist in organizational cultures, we can see that there may be a problem with the manner in which we communicate with one another. Both written and verbal communications are necessary to ensure effective communication.

It was during a leadership training program that a discussion came up regarding rules, operating guides, and procedures. A participant was complaining that his department had too many rules. "We've got too many guidelines," he said. Let's examine what would happen if we didn't have rules, or guidelines, to follow. You can readily see that it would encourage freelancing because each member would be free to do his or her own thing based on their perception of the problem and its probable solution. The resulting behaviors may or may not be in concert with what the other members would agree on.

Standard operating procedures and standard operating guides provide the basis for the uniform behavior of all members working at an incident. Working under emergency conditions is a dangerous proposition at best and gets worse as time goes on. It would be irresponsible to arrive at an emergency incident without a game plan. There are not many scenes more chaotic than an emergency scene. The emergency personnel are expected to bring some order to the chaos that they find on their arrival. Without an initial set of procedures to direct their behaviors, they would run the risk of contributing to the chaos instead of reducing it. Emergency services supervisors have the enormous responsibility of caring for both the safety of the public and the safety of their own subordinates. Their greatest asset is their knowledge of the procedures and guidelines. Well thought out, tested, and clearly written, standard operating procedures and guidelines are tools that empower members to carry out their duties.

A young female firefighter was assigned as a driver for a deputy district chief for the first time. Part of her duties required her to walk around the fire scene and identify the location of the working units then report back to the command center. The command staff would use her information to chart the units and their locations on the tactical board. The incident commander would use this information to direct the extinguishment of the fire. After observing the command staff's communication with the sector chiefs to get regular progress reports and the identities of the companies they were supervising, the young firefighter said in amazement, "Wow! You guys know where everybody is working here." She went on to say that she "never knew that there was so much support given to members working at a fire. Knowing this will make me feel safer when I am working in a fire building."

The incident command procedure that this firefighter witnessed was standard operating procedure for the Chicago Fire Department. She had heard of it before, but this was the first opportunity that she had to see it in operation from the inside and to be a part of it. The objective of the incident command procedure is to keep track of the units and the members assigned to the incident.

All standard operating procedures and guidelines should be evaluated regularly to determine whether they continue to be useful. Operating on an emergency scene without guidelines or procedures would create situations that could entrap the task providers in a web of inconsistent direction and conflicting behaviors.

There are a number of fire departments that have very little if any disciplinary procedures; instead, they operate by force of personality. Discipline is dispensed according to who one knows or who they are. The people who are liked by authority figures often get favorable treatment, while those who are not liked receive harsher treatment for similar offenses. Supervisors are sometimes put into a position in which they have to verbally plead with subordinates to get them to behave properly. It is difficult to enforce unwritten rules equitably because too much depends on personalities. This uneven application of the rules creates an atmosphere of distrust and misunderstanding that entraps workers in a state of confusion by sending them mixed signals. A thoroughly written discipline procedure that is evenly enforced empowers members by giving them clear descriptions of unacceptable behavior. Some members have reported an increased sense of personal security when working under such clear and objective rules.

Labor contracts are legal documents fully enforceable in a court of law. I was recently asked to conduct a leadership workshop for a fire department in a medium-sized city in central Illinois. This department had been experiencing labor problems for a long time, and I was brought in to try and ease some of the tensions. At the very beginning, I informed the group of chief officers that they should use their department rules, regulations, and labor contracts as administrative tools. One chief asked if that meant that they would have to follow the contract. I explained that failure to follow a labor contract could result in the city having to defend itself in a court of law or having to go before an arbitrator if any violations of the contract occurred. To my surprise, no one had ever explained that to them. Even more surprising was the fact that the labor union had not taken the city to court for ongoing violations of the contract.

Labor contracts are very helpful for supervisors because they describe management rights as well as worker's rights. The contract can provide a guide to work by. On many occasions, I have referred members to their labor contract to see if it permitted me to fulfill a request that they had made of me. It is not uncommon for members to ask for privileges that are prohibited by their labor contract. Supervisors should be careful because a contract violation, even at the request of a protected member of the labor organization, can lead to an official grievance by the union. Such a grievance could result in a monetary award in favor of the union.

"At what point do we stop empowering people and start telling them to do the work?" This was a question asked in one of our workshops by a chief who went on to explain that the officers in his department had received training that encouraged them to empower their task providers. Empowerment had been defined to the officers as giving their people more responsibilities and input into the overall operation of the organization. There had been no mention, however, of additional training to prepare them for the additional responsibilities. The chief agreed that the lack of training could be a reason that the new empowerment effort in his organization had stalled. The lack of effective training has been identified as a reason why supervisors don't delegate more. Supervisors are hesitant to delegate work to workers who have not demonstrated an ability to do the work. Effective training could be the solution for both of these concerns.

The definition of empowerment in the context it was presented to the officers was very close to the definition of delegation of duties. Empowerment, as I intend it, will place the emphasis not only on what to do, but also on how to perform the duties required. Training is an important element of a supervisor's responsibilities to his task providers. In addition to preparing and presenting knowledge, effective training programs will require some action on the part of the learner. Supervisors and instructors generally do a fine job until they get to the application step of the training session. This step requires that each of the class participants demonstrate their new knowledge or skills. Instructors are often reluctant to put the learner on the spot because they are concerned about embarrassing them. I don't know of any other way to have the learner demonstrate their competence in the skill or knowledge that has been presented to them. As learners, if we can't tell it, or if we can't show it, we don't know it.

When you put a person on the spot, it is not to cause them to fail; it is to give them an opportunity to demonstrate what they know and to demonstrate their level of competence. This is a point at which they can earn some positive recognition for a job well done. It is very important that when an instructor or supervisor puts someone on the spot, they understand that that person is in their protective custody. You are not to allow any harm to come to this individual because you want to give him/her an opportunity to demonstrate what they know. The trainee is not there to be ridiculed by other members. So it's okay to put a person on the spot, but you must also protect them. One of the things I do when I put a person on the spot is have everyone in the room give them a big round of applause when they finish their activity. As trivial as that may sound, it not only benefits the person receiving the applause, but it also brings a positive aura to the learning environment. As human beings, positive recognition and applause encourage us. This is done to make the participant feel good about completing the exercise.

If a participant makes a mistake, we get right on the spot with them and make suggestions; for example, saying something like, "Let's see what happens if we try it another way." We assist him/her to be successful. This should be done without being offensive. The objective of training is to pass on knowledge and skills to empower the task providers to successfully meet the challenges of their tasks. Once the task provider has demonstrated a certain level of competence, we have a basis for expecting them to successfully handle the duties that come with their position. We must let our people know that their success is important to us. Until their level of competence is determined, we don't know when there is a discipline problem. The worker simply may not have the knowledge or the skills to do the task. Once the skill level is determined, the organization can hold its members accountable for disciplinary procedures.

I think that the following quotation may provide some insight into the central problem of training adults.

> *Every act of conscious learning requires the willingness to suffer an injury to ones self esteem. That is why young children before they are aware of their self-importance, learn so easily; and why older persons, especially if vain or self important, have great difficulty learning at all. Pride and vanity can be greater obstacles to learning than stupidity.*
>
> *~The 2nd Sin~*

As a district chief, when I got a report that there was a problem in a firehouse, without exception, my first step was to refer to the labor contract, department policies, and practices and procedures to make sure that they were being adhered to. In one case, an officer told me, "No, I don't go by the rulebook, or the labor contract. If a firefighter comes to work five minutes late I overlook it, but I don't want him playing basketball in the afternoon. That's my trade-off." He went on to explain that the members of his company liked to play basketball in the afternoon while he was trying to sleep. That was a problem because he thought that if he was overlooking their violations of the rules and department polices, the very least the firefighters could do in return was let him sleep peacefully in the afternoon. The spokesman for the other side of the situation that involved a White officer supervising a company of Black firefighters spoke up and said, "The officer won't let us play basketball. You know how we are, as Black people we like to play basketball." He reasoned that since I am Black, I should understand his position. The Black firefighters saw the conflict as an ethnic problem. I told the group that the Chicago Fire Department does not pay people to play basketball or to sleep in the afternoon. Playing basketball is a privilege not a right. Sleeping in the afternoon and not following the directions of the rulebook are violations of department policies as well.

My challenge at that point was to get the company back together and get the Captain back in control of his company with a minimum amount of trauma. The officers were directed to have their members take down the basketball backboard. While that was being completed, I convened a meeting of the officers that included the deputy district chief, the battalion chief, and the two company officers of the companies assigned to the firehouse. The officers were informed that there were clear violations involved, but because re-establishing the captain's control of the company was more important than pursuing remedies through discipline procedures, at that point, I would not begin discipline proceedings against him. It was explained to the captain that when an officer ignores department polices and the labor contract, he throws away the source of his authority. Without them, the officers must rely on the force of their personalities and pleadings for cooperation from their subordinates.

After the officers' meeting, we joined the firefighters in a joint meeting where the new operational procedures for the company were described. The basketball backboard was not to be put back up without my permission. These new procedures included twice-daily visits to the firehouse by the battalion chief at the morning roll call and afternoon training sessions. They were directed to document all violations of department polices and forward the reports to my office without delay. The officers were informed that their failure to keep accurate records and to enforce the department's rules, regulations, practices, and labor contract would result in discipline charges being placed against them as well as the offending members. The group was assured that my commitment is to protect them when they are right and correct them when they are wrong. It was understood that we were not going to be unreasonable, but that unacceptable behavior would have consequences. The captain regained control of his company and the members found that they could live quite well with rules and regulations and the equitable enforcement of them. Several of the members involved in this scenario have since been promoted.

When first informed of this situation, I was told that there were the makings of a race problem in the firehouse. The officers and the battalion chiefs were concerned that the intensity would get out of hand. Reconciliation meetings conducted by the deputy district chief and the battalion chief with the involved parties had been unsuccessful. The situation became worse when a company officer's fireground radio was reported stolen and some damage was done to the captain's personal vehicle. As the officers came in with the police reports, they expressed their frustration and fear that something bad was going to happen in that firehouse. The firefighters having heard that the officers were coming to see me asked for and were granted permission to meet with me. Both sides expressed their fears that the situation would get worse before it got better. The firefighters were concerned that if things continued on without some intervention, they would be the ones hurt most. Both sides were caught in a circumstance that they could not see their way out of. All of the parties thought that this was a racial conflict because the firefighters were Black and the officers were White. In addition, there were some White firefighters who did not get into the conflict, but rather remained neutral throughout the entire ordeal. Basing most of my decision-making on the department's administrative documents helped minimize the issues of race. The interests of all of the parties were served with a minimum amount of embarrassment because everyone was encouraged to play by one set of rules. This resulted in a win-win for both sides of the conflict as well as for the community that we serve.

The fact is, the conflict was over privileges and to whom they are entitled. Much of what we have conflicts about in our society has to do with privileges, either earned or unearned. In this situation, the officer granted the unearned and illegal privilege of breaking the rules in return for the unearned and illegal privilege of sleeping in the afternoon. As administrators, it is our duty to enforce the rules and regulations of our organizations. I have been surprised at the number of difficult situations that have been resolved or prevented by the equitable application of the organization's polices and practices. Once the rules and polices are put into place and evenly enforced, many of the racial, gender, and cultural issues are minimized. In the absence of the rules and polices, racial and gender issues are maximized. In this case, a failure to empower the members using department administrative documents would have resulted in a maze of confusion and misperception. It is worth noting that people of different races and genders can have conflicts based on issues that are not racial or sexual in nature. In this case, the concern for racial differences blurred the more important rule violations.

Chapter Five

TRAGEDY ~OF THE~ AGE

Herein lies the tragedy of the age: not that men are poor—all men know something of poverty; not that men are wicked—who is good? Not that men are ignorant—what is truth? Nay, but that men know so little of men.

~W.E.B. DuBois~
"The Souls of Black Folk" (1903)

CLEARLY, THE TRAGEDY OF THE AGE is that human beings know so little about relating to one another effectively. The result is a world filled with interpersonal chaos. I would like to look at personal conflicts, regardless of their nature, from a generically human perspective, rather than from a race or gender perspective. All human conflicts have a human element that for some reason is often overlooked, especially in the early stages of conflict. In many instances, the fact that we are all human beings is considered only after hurt and bloodshed have occurred.

If we look at a group of horses, we would not say that a black horse is superior or inferior to a white horse or a brown horse. We see horses, and we admire them for their physical beauty and other natural characteristics. I would think that the color of their coats, while obvious, would be of less interest.

When we look at other people and ourselves, far too many of us are concerned with race and gender. We often determine who is worthy of respect or who is competent based on their color. The assumptions that White people are naturally more competent than Black people and that males are naturally more competent than females are simply erroneous. Competence comes from preparation, training, and education; one's race or gender does not predetermine it.

Derrick Bell, in his book, *Faces at the Bottom of the Well,* presents the premise that racism in America is permanent, that Blacks will not overcome it, and that overcoming racism in America is a myth. Although he ultimately says that he is outraged by such positions, some of his readers probably wonder, "What are we to do if this is true? Why continue our efforts to overcome the ravages of racism and bigotry? Should we view this as a rejection of our humanity, and should we self-destruct?" For Black readers who have found themselves reacting this way to Bell's writings, I advocate a change in self-perception. Instead of seeing ourselves as people whose empowerment is doled out to us by others as entitlements owed because of slavery, we should recognize and embrace the fact that we are actually self-empowering. We do not need entitlements bestowed upon us by others; we can use our natural gifts, our endowments, to assume the responsibility for empowering ourselves individually.

The four endowments that each one of us is born with are self-awareness, imagination, freedom of choice, and independent will. The recognition and effective use of these endowments is a good place to begin leveling the playing field of interpersonal relationships.

An insult from someone who is perceived as superior seems more humiliating than one from a person seen as equal or inferior. An effective response to an insult may be for the target to consider the source of the insult and dismiss it. This kind of healthy response is much easier to give when one has a sense of empowerment, rather than when one feels powerless and vulnerable to the judgments of others.

Bell states in his epilogue that there are at least two things that have led to Blacks being successful, a strong will and an imagination. Bell's words reminded me of slavery scenes in which enslaved Blacks in cotton fields were working in the hot sun, singing spirituals, and pushing on. They survived because of their will to get through the day and their ability to imagine a better time to come.

As I advance the concept of human endowments as the basis of personal empowerment, I will elaborate on the four endowments previously mentioned. How we use them will determine whether we succeed or fail in our attempts to relate effectively to others. In order to succeed, we must first make a commitment to be responsible for ourselves.

In his psychological research, Dr. Victor Frankl identified two distinct kinds of people: those who are responsible and those who are irresponsible. I believe we choose to be either one or the other. While we have the right to choose to be irresponsible, we also have the obligation to be accountable for our behavior.

I propose that as human beings we have the right and the ability to revoke the authority we may have given others over our existence. If we do so, we are choosing to assume responsibility for our personal behavior. Many conflicts are based on one person attempting to exercise control over another without recognizing that the other person must cooperate. We must understand that this cooperation is revocable and conditional. In many cases, the "victim" chooses to put the responsibility for his or her own well-being into the hands of another party, but the other party may not want the additional responsibility. It could be reasoned that it is an imposition to charge someone else with the responsibility for our own well-being. It is possible that the conflicts in which we often find ourselves stem from the fact that every human being has the full-time job of managing his or her own endowments. We may resent it when others overload us with theirs.

Furthermore, we behave according to our personal perceptions. If we perceive ourselves as powerless, then we will assume the behavior of the powerless. The perception of being powerless, of being a "less than" in society, leads to a processing based on surrender and defeat. The novelist, Ernest J. Gaines, whom we will discuss shortly, poignantly portrays this process. The perception that we have of ourselves has a much stronger impact on our behavior than any perceptions that others may have of us. When we perceive ourselves as being adequate, we behave in accordance with that perception. We move with less stress and increased confidence. If we fail to make responsible choices for ourselves, we will be directed and controlled by those who do.

Most importantly, we must understand that there are choices. We can create and fulfill our own objectives, or we can allow someone else to create objectives for us to fulfill. Individuals often join gangs or mobs to avoid making responsible choices, thereby enslaving themselves to the leaders of the group. If the objectives of the group do not support our personal goals, we should reconsider our support of that group. We cannot make that determination without identifying our personal goals and then comparing them with the goals of the group. If the two sets of goals are consistent with one another, then joining the group could be mutually beneficial.

In *A Lesson Before Dying,* Ernest J. Gaines tells the story of a young Black man living in Louisiana in the 1940s who is on trial for murder. In the small Cajun community, injustice and cruelty toward Blacks still prevails nearly a century after the abolishment of slavery. Having been brought up believing the myths that all Black men are animalistic, unintelligent, and sexually deviant, the White defense attorney characterizes the young Black man, Jefferson, as a "mindless hog who should not be held responsible for his actions". The attorney tells the jury, "I would just as soon put a hog in the electric chair as this." Jefferson, whose spirit is broken, accepts and thereby internalizes his lawyer's dehumanized depiction of him.

As the trial comes to a close, Jefferson is convicted for a crime he has not committed and is given the death sentence. Jefferson's grandmother, Miss Emma, extremely distraught by the wrongful verdict, is even more heart-broken that her grandson's humanity and dignity have been taken away from him. Seeing that Jefferson has no strength or will to resist his fate, Miss Emma solicits the services of a college-trained Black teacher, Grant Wiggins, and begs him to teach Jefferson how to regain his human dignity, a lesson that will allow Jefferson to die like a man.

Gaines movingly describes the difficult transition that Jefferson goes through as Grant tries to instill in him a sense of pride and worth against the seemingly hopeless backdrop of death row. Part of this process includes having Jefferson write about himself in a diary. With this act, Jefferson is able to describe his perceptions of himself and eventually share his feelings about his grandmother as well as other people who had shown him compassion. Interestingly, he begins to return acts of kindness and to embrace and appreciate the concern that people in the community have expressed for him. Gaines allows us to watch Grant help the resistant Jefferson see that he does not have to accept the negative and degrading images of himself that have been imposed upon his psyche by a suppressive culture that refuses to acknowledge him as a man, a human being. After months of positive reinforcement, Jefferson begins to see himself as a man—a man who knows he has to be strong for his grandmother as he faces his impending death. Eventually Jefferson realizes that although he cannot escape his inevitable execution, there is beauty and heroism in resisting it. On the day that Jefferson is put to death, he displays the fortitude and courage to die with dignity.

Chapter Six

INTERPERSONAL COMMUNICATIONS

AT THE BEGINNING of my fire officer workshops, I like to have each of the participants come up to the blackboard and write down a concern that they have regarding the topic of discussion. These concerns should be anything that they are uncomfortable dealing with or they do not feel confident addressing with their subordinates. It is important that the leader be the most secure member of their command. If not, insecurity will trickle down to the members of the command.

There are several other reasons for having the members express themselves in this way. Most importantly, each of them will have made a contribution to the learning experience. They will have taken a risk by presenting themselves and their concerns before the entire class. This encourages active participation. By using these concerns as class objectives, we will discuss items that they are interested in, rather than discussing only those items that I would hope they'd be interested in. After the concerns are listed, I ask each writer to give the class some clarification of the circumstance or background of their concern. Each of their concerns provides a teachable moment for me to present information in the context of their interest. The result is effective two-way communication with adults talking to one another, rather than a situation in which the instructor talks "at" the students. However, as the instructor

I have the additional responsibility of providing for their comfort, maintaining classroom decorum, and giving them timely breaks.

Classroom participation is the basis for a feeling of ownership in the students' learning experience. The key to success for this technique is strong support from the instructor for the student population. The students must feel assured that their instructor will not allow them to be subjected to personal ridicule. Here is a sample of some of the concerns and my responses to them.

Affirmative action makes me uncomfortable as a company officer of a growing department.

The speaker went on to say that this is of greater concern now that a growing number of minorities and women are being hired. My first question is: "Is there anyone in this group who does not think that all citizens should have an opportunity to fail?" The answer to this question is pivotal. I have yet to get a negative answer to it.

The next question is: "How does the opportunity to fail differ from the opportunity to succeed?" The answer is that they are the same. As leaders we are expected to have an interest in providing opportunity for each of our subordinates to succeed. The objective of each of these laws is to give opportunity to the protected groups.

The question then becomes: "Can programmed opportunity be given to someone without others being inconvenienced or made uncomfortable?" Obviously the answer is no: there must be a price paid. In order to be cost effective, the price should not be any more than is absolutely necessary. The result should be a stronger nation with a larger pool of solution providers. The intrinsic values of increased harmony and self-reliance will be immeasurable. Those who object to the inclusion of minorities and women into the mainstream say that they cannot compete and are sure to fail. An effective affirmative action program should render itself unnecessary upon success. When women and minority members can fail and it is accepted the same way as failure within the dominant community, we will know that affirmative action has succeeded. Currently, when a member of a minority community fails, it is treated as an indication that the entire community has failed. When a member of the ruling community fails, it is viewed as an individual's failure, not their community's. Similarly, some believe that when a woman fails all women have failed. A common statement is: "They're all like that."

When one is denied an opportunity or accommodation because of their race or gender, it is an attack on their basic dignity. Just as in *A Lesson Before Dying,* the issues have to do with human dignity, and some people would rather die with it than live without it. Dignity comes with birth based on the gifts of imagination, self-awareness, freedom of choice, and independent-will, and the responsibility to use them in an effective manner. When we recognize these elements in ourselves, it is easier to see them in others. Then basic dignity becomes something that one can only lose through irresponsible behavior, instead of something that one has to earn from another human being. No one should be put into a position requiring him or her to convince others that they are worthy of basic dignity and respect. That is a birthright.

Irresponsible behavior must be disciplined. The objective of the discipline should be to adjust the unacceptable and/or irresponsible behavior, not to challenge the person's dignity. Leaders must learn to separate a person's behavior from their value as human beings. An individual's behavior is to be evaluated, but a human being's value should not to be judged. Most people are innately good, even though some of them behave badly. As we learn to separate their behavior from their humanness, we will be able to adjust their behavior without damaging their personal dignity. People find it less threatening when their behavior is challenged than when their dignity is challenged. It is as if our behavior is in the public domain subject to evaluation and/or review. Most of us realize that we can change our behavior. Basic dignity on the other hand is not subject to the same kind of evaluation or review. It is difficult to resolve issues involving an individual's dignity. I am reminded of a group of inmates in a Pennsylvania penal institution that pooled together $1,700 to contribute to the New York World Trade Center survivor's fund from their 19 cents per hour earnings. When asked why they did it, they explained that even though they are convicts they are also human beings. This is an example of the dignity that remains in human beings who have behaved badly.

What are acceptable or unacceptable accommodations to cultural diversity?

In order to understand diversity we need not go any farther than our own families. No two children have the same two parents because no two children have the same needs. The sensitive parent being aware of this may serve a different fruit to each child based on his/her taste. It may be that one child may prefer an apple while the other would prefer an orange. In a like manner, an organization may accommodate the preferences of distinctive individuals or groups. Where the law requires these accommodations, they must be made. In many instances these accommodations can be made with minimal inconvenience to other workers or the organization. This may encourage demands for similar treatment from other groups or individuals. Like children in a family, other members will ask for an orange because another member got an orange, even though they like apples better. Further discussion may reveal their preference for apples once they are informed that they have a choice.

Leaders should not be disturbed by these demands. On the contrary, leaders should be prepared to expedite them. Assure the members that you will forward their written request to the appropriate party and get back to them with any additional information that you receive. Leaders are to give their subordinates' concerns their full consideration. It has served me well to assist subordinates in the preparation of their requests. These are rare opportunities for positive interaction with them centered on their interest. Having them submit their request in writing tends to sort out the trivial complaints from the serious ones. When discussing racial, cultural, and gender demands, be sensitive to the basic human dignity of those involved, being careful not to injure it unnecessarily. Remember that at the base of most diversity concerns are human dignity issues.

It is not beneficial for a leader to have their subordinates think that he/she has more authority than they actually have. Often that will encourage requests for unacceptable or unreasonable accommodations. Subordinates who think that the supervisor has more authority than they actually have may attempt to ingratiate themselves with the supervisor to gain undeserved favors. One such favor may be to accept verbal requests instead of written ones. A typical assertion might be, "Because I bring you hot coffee every morning, I should not have to put my requests in writing." This type of relationship should be avoided whenever possible. Beware of those members who try to ingratiate themselves with you. The favors that they offer and your special treatment of them can undermine your relationships with the other members of your command. In my experience, I have found that ingrates are often the least competent members of the team. They ingratiate themselves as an attempt to cover their incompetence. The common good is best served when they are trained and held to the same standards as the other members of their company, and supervisors should get their own coffee.

A member's reluctance to put requests in writing may open you up to trivial requests, and the person may also blame you personally if their requests are rejected. Writing a request provides an opportunity for the writer to clarify and crystallize it. To minimize personal resentment towards you from members of your command, offer timely and objective reasons for request denials. It should be established early on in your relations with them that while you will accommodate them as best you can, the rules, regulations, department policy, and labor contract will be strictly adhered to. So, while we all have the same human dignity, it is not uncommon for us to have diverse interests and preferences. The challenge is for us to value our differences, not be conflicted by them.

Some members of my command are bitter after being passed over for promotion because of our department's affirmative action program. How do I handle that problem?

Certainly, we can understand his rage after a member has done all that they can to be promoted only to have it denied them when they come up for promotion. This rage should not be trivialized in any manner. I am reminded of a diversity class I was leading for a group of paramedic recruits in Chicago. One of them stated that he was against the quotas used to implement affirmative action. He made a case for those people who were not hired in rank order because minorities and women were being hired out of rank order before them. After he finished his accusation, he was asked whether he knew that in addition to the quotas he mentioned there is also one for paramedics who are on the existing firefighter's list. He said that he was not aware of such a quota. He went on to say that he was on the firefighter's list. After being informed that he qualified and could be hired as a firefighter in the next hiring call, he said that quotas were not as bad as he first thought. His position in the rage/guilt syndrome was reversed.

In Chicago, the federal courts directed the city to develop an affirmative action plan. The plan permitted a percentage of all hiring and promotions to include minorities and women out of rank order. Any White males not hired in rank order would be placed at the top of the next hire or promotion list. These people are called passovers for the obvious reason. Their promotion or hiring might be delayed, but not denied.

One of the questions that I often hear is: *"How do I recognize a bias in myself?"*

First of all, I don't know of anyone who does not have a bias. I think that it is human to have personal preferences. In the workplace however it is unacceptable to make decisions based of those preferences. Professional decisions should be made on sound objective and definitive facts. Failure to do so may cause you to be on the losing end of some future legal action. A Black police commander once asked me if I thought a White supervisor who worked for him was racist because he was friendlier with his White subordinates than with his Black subordinates. There were no problems with work assignments or any other work-related issues. He said he did not think that it was racist behavior because he tended to socialize with his Black subordinates in a similar manner. "Exactly what is racism anyway?" he asked.

I answered him this way: "Any decision or action based exclusively on racial or sexual concerns is racist or sexist." In the workplace it is unlawful to do so. In the workplace, leaders should make decisions and take actions based on definitive and objective deliberations. The references for these deliberations should include the organization's policies and labor contract. I don't know that having a cup of coffee or tea with members of your race or sex is a violation of any laws or policies. If the supervisor held back earned recognition for their work, they would have a valid issue, especially if others were given recognition for similar work.

I wonder if the people who were complaining would feel the same way if they valued their innate human endowments as the source of their basic dignity and not the friendship of their supervisor. It would not surprise me if they were to feel differently with the increased self-esteem that comes with the recognition of this fact. When one understands that they are to use their human endowments to become all that they can be they tend to be less threatened by others also trying to be all that they can be. These innate gifts will contribute more to their success or failure than any other elements of their existence.

I am having a really difficult time supervising the members of my company because they are my friends and fishing buddies. I tend to overlook their faults and would rather do their work for them than to get into a hassle with them.

It may be helpful to you if I share one of my experiences with you. I was assigned to a hook and ladder as a new lieutenant. The members of the company had been friends of mine for over 20 years. My driver had taught me how to use forcible-entry tools such as axes, pike poles, and pry bars as a recruit over 20 years earlier. Others had been recruits with me, so you can see that I was in a situation similar to yours. One of the first confrontations involved getting them to do training drills. The first couple of times they would say, "You know that we know how to do that already." My response to them was, "That's great because then it will only take us a few minutes and then I can get back to my paper work." Or I would say something such as: "I know that you guys know all about this (whatever the subject was), but I need a refresher." I would then invite them to teach me what they knew about the subject of discussion.

The company officer does not have to lead all of the training sessions. There are members of your company who have expertise that they would be glad to share with the other members of the company. This provides them an opportunity to earn positive recognition, which they enjoy (see the chapter on Human Behavior). It was interesting that once we started the exercise, it would usually take longer than we expected because one question would lead to another. The result was an effective training session.

It is important that the training be relevant to the members. A brief discussion explaining the importance of the subject helps to stimulate interest in the training. After a short while, there will be less resistance to training. With my old friends, it helped that I went over our job description with them and asked them to assist me in meeting my responsibilities. That was my way of respecting their dignity and expecting that they would respect mine in return.

As officers we have to ask ourselves whether it is more important to be liked or more important to be respected. We should understand that with each of these answers, we have a different set of behaviors. I am not here to tell you how to act—that is your choice—but your behavior should be consistent with your desired objective. I will say to you that one can behave in a manner that will earn them respect without being argumentative or losing friends. The key to gaining another's respect is to give them respect. When a leader demonstrates an interest in the welfare of his/her task providers, they have made great progress toward gaining their respect.

Another major step toward developing respect is for a leader to demonstrate interest in the welfare of their task providers. The leader who can demonstrate how skills learned on the job can benefit task providers in their personal lives will have less trouble getting cooperation from them. It is very important in this situation that you let your members know that it is your desire to assist them to get promoted and improve the quality of their lives. I have a problem with supervisors who don't think of their subordinates as friends. Who are better friends to you than those people who help you meet your responsibilities in the workplace? They are in fact major contributors to your welfare.

After going over our job descriptions, the members of my company were surprised that so much responsibility goes with the supervisor's position. You may be pleasantly surprised to find that the members of your company will be more than happy to do their jobs once they realize how important they are to the overall success of the company. True friends help one another when given the opportunity. Let them in on the decision-making. Tell them what the problem is and ask for their input. It may be that you are doing their work for them and they do not realize that it is their responsibility. They should at least be informed of their duties even if you don't enforce them. A favor is not a favor if the person receiving it does not know it to be one.

Many of our people do not know what department rules and regulations require of them. Much of what we do in our day-to-day activities in the firehouse has little to do with what the rules and regulations call for. Many times I have conducted training classes on department policies and practices and have had veteran department members express amazement by the number of duties that the policies required of them. Much of the success on promotional examinations comes from a thorough knowledge of the department's rules and regulations as they are written.

In your effort to preserve the friendships of your subordinates, you may in fact be doing them a disservice. In closing I will share this story with you. The ringleader leading the opposition against my first training exercise later informed a group gathered at a birthday party that "Bennie Crane was the only officer that ever tried to teach him something." What he did not say was that he always challenged the officer's resolve to train.

There is no support from the duty chief during drills.

A lieutenant explained to the group that his immediate supervisor, the duty chief, does not support him when he is conducting his daily training sessions. If the chief is in the firehouse when he begins his training, he leaves. The speaker added that, "It would be helpful to me if he were to at least observe the training and give me some moral support occasionally."

In response to his concern, I asked the lieutenant if he had ever considered that perhaps the chief was not confident of his own knowledge levels, and did not want to risk embarrassment. I have found that to be a problem with some chiefs. I always like to see chief officers in my classes because it tells me that they are interested in learning and sharing their knowledge with others. It does take some courage for chief officers to attend training sessions with their subordinates. On the other hand, leaving the training completely up to you could be an expression of his confidence in your skills. Believe me, in some cases, it is best if the chief does not attend your training sessions because their presence can be a distraction.

There is this one firefighter assigned to my company that doesn't get along with any of his fellow firefighters.

I asked exactly what it is that his co-workers dislike. His response was that, "He just does not fit into the group." The captain said that he could not explain it in clear definitive terms. I told the captain that if he can't explain the problem in clear objective terms, chances are it is an attitude problem. I went on to tell him that if it were an attitude problem, he should be very careful in addressing it. If the person does not violate any rules or regulations and does his work in an appropriate manner, there is not much that a supervisor can do. On a personal level, a supervisor can meet with the individual unofficially and discuss the concerns, being very careful not to tread too deeply into their private life. Some time ago, it was brought to my attention that most Americans live in quiet desperation. I think it is very important that supervisors let their people know that they are interested in their well-being. We just don't know what kind of problems our members are experiencing at a given time. I can't tell you how to do it, but it starts with having a commitment to help and watching for the opportunity to do so in a positive manner. Let me share the following with you.

I had 650 firefighters and officers assigned to my district—far too many people for one person to have a personal relationship with. As part of my concern for them, I sent out birthday cards to each one of them with a note of appreciation for their assistance in helping me provide the best possible emergency services to the citizens that resided in our fire district. In return I often got expressions of appreciation. One of my members told me that his birthday card had been especially important to him that year because his mother had died recently and he was the only one left in his family. You may be surprised by the amount of good you can do by just having a cup of coffee with your people.

Our chief holds our entire group responsible for the problems that he may have with a member of our group. I don't think that is right. What do you think?

I have never agreed with that kind of discipline, not even in grade school. Aside from that, I don't think that it is legal today especially if there is a labor contract involved. Every adult should be accountable for his or her behavior. The supervisor operating in that fashion is creating a platform for his subordinates to unite against him. This has the sound of an operation being run by force of personality. That is, when one of you makes a mistake, he doesn't like any of you. I can't imagine any rules or regulations that would allow for that kind of behavior. It disrespects the basic dignity of those responsible members caught up in the situation. That sounds like an administrative disaster waiting to happen.

We had a similar situation that involved a lieutenant who referred to his people as "You people" and "Hey you". He completely disregarded each person as an individual. This situation came to a conclusion one day when he ordered "Hey you" to fill a hand pump and nobody moved. He was then informed that "Hey you" didn't work there any more. When the situation was brought to the district chief's attention, the lieutenant was directed to address his subordinates by their rank and proper names. Again, it is very important that basic human dignities be respected at all times. Your chief would have a hard time defending his group discipline before an arbitrator.

The staff officers of our department do not participate in the department's mandatory drills.

First we should determine if their participation is required by department policy. If department policy directs them to participate, their failure to do so constitutes a violation and should be addressed as such. I don't know of many staff officers who drill on a regular basis. They are usually exempted from routine training because specialized training may be required of them for their staff responsibilities.

In our firehouse everybody has their own agenda. There are a couple of people that are always complaining about one thing or another. Sometimes it seems as if they thrive on turmoil.

You should know that there is nothing that says that we have to be friendly everyday and get along with each other. This is a job, not a social club. Again, this is the reason we have department policies and procedures to follow. The best way that I have found to control petty complaints is to have the complainer develop his own solutions to his complaints. As a supervisor, I did not accept complaints without a recommended solution. Many times people complain just to be doing something; it might have something to do with their nerves on a given day. Another reason could be that the members don't have enough to do. That is, they may not be appropriately challenged. I have spent most of my time in the fire service in busy areas. I can certainly imagine that it would be nerve-wracking to work in slower areas. When I talk with members who work on slower companies, I advise them to take advantage of the time to study and sharpen their skills.

Because we live in close proximity to one another, I am not surprised that we have personality conflicts. In fact, I am surprised that we don't have more conflicts. We spend more time with each other than we do with our families. It is for that reason that I have included the sections on human behavior and conflict resolution. The information found there should help you understand and manage personality conflicts.

My shift commander has his mind set and won't even consider anybody else's ideas. He is not a bad guy and he is right more times than he is not, but the times that he is not it is difficult to deal with him.

Well the boss is not always right, but he is always the boss. This is one of the tougher situations that I hear about in these workshops.

Some of the techniques that are discussed in the Human Behavior chapter of this text should be helpful. After reading the material found there, you should have a better understanding of his behavior. I just wonder how he would respond to some information that would get him additional respect and recognition without someone taking credit for showing him how to get it. That is the secret, lay the idea out and let him take the credit for it. It requires a very secure person to pull that off, but it has been done successfully in situations where the common good was given priority.

There is a minority that threatens to file an affirmative action complaint whenever his officer holds him responsible for doing work around the firehouse. He seems to think that he can choose the work that he wants to do or not do. As his officer, I don't want to confront him because I don't need the grief that is sure to follow. What do you suggest I can do without creating trouble for the department?

This is a recurring problem today because we live in a society where lawsuits are a growth industry. People are suing one another in record numbers and winning large amounts of money. As fire officers, we can be sued individually if it can be proven that the individual failed to follow department policies. I have had several complaints filed against me because I held minority members accountable and responsible for the duties outlined in their job description. It is almost impossible to maintain the integrity of your command if you allow some members to do their own thing while holding other members to department policy. As supervisors we have no other options.

One of the first things I would do is make sure that he is thoroughly familiar with his job description. That would be the first question that an arbitrator or a judge would ask you. You would not believe how many cases have been lost because the member did not know his job description. The courts have often decided against employers who could not certify that their members knew what was expected of them or that they were adequately trained. This is a major problem in American businesses today. Managers don't tell workers what is expected of them and how to do it.

After that is done, you will need to document all of the individual's activities. This is a part of your administrative responsibilities. This will save you a lot of time and grief later. I have yet to lose a case. My success has been due to the maintenance of appropriate documentation. I have been retired for six years and have numerous files of complaints that were brought against me during my career. Be careful not to get caught up in the emotional whirlwind of a case. Don't take the charges personally, even if they are. If you work within the guidelines of your department policies, you should be protected as an individual from lawsuits.

Chapter Seven

TEN 7-TON *ELEPHANTS* *STANDING IN A ROW*

IMAGINE…ten 7-ton elephants standing in a row. Each elephant is connected to a cable attached to a large shackle around its right hind leg. The other end of the cable is attached to a small stake driven a few inches into the ground. Certainly those shackles and cable are not enough to hold these elephants in place should they decide to walk away. Why are they standing in a perfect row? What holds them in place? The explanation is simple. According to the story, when they were young calves, they were connected to a very heavy chain that was attached to shackles around each of their right hind legs. The other end of the chain was connected to a very large stake driven several feet into the ground. As young calves, they would struggle for freedom, usually until they were too exhausted to continue. After a while, they realized that they were not going to be able to break away and that it was easier to stand still than it was to struggle. The calves became conditioned to respond to the shackles and cables by staying put. That learned behavior carried over into their adulthood.

In many instances, people are vulnerable to the same type of conditioning. Some of us allow history and past experiences to provide the basis for our current behavior and attitudes. The experience of being brought to America in slavery functions as shackles and cables for African-Americans, keeping us in line, holding us in place, and stymieing our creativity rather than providing a basis of knowledge and experience on which to build.

It is not uncommon to hear sentiments such as, "We've never been able to do that", or "If it was good enough for my parents, it's good enough for me." Statements like these are rationalizations made by people who are trying to justify staying in line, while being held in place by outdated information and concepts. Just as the elephants, who because of their past experiences as calves, continue to behave submissively, many of us, too, behave like compliant children or subordinates, not only because of our own past experiences, but also because of the experiences of others that we perceive as similar to our own.

This kind of thinking provides the basis for one to assume a posture of irresponsibility, the implication being we cannot be held responsible for the direction our lives take because we lack the power to change our circumstances. So you see, history can provide a rationale for staying in line and accepting the status quo. But to do so is a choice. People choose to be powerless. People choose to relinquish responsibility for themselves and allow others to direct their paths.

Those of you who find yourselves feeling powerless because of past conditioning should consider the paradigm of "the half-life of knowledge," which states that half of what we learn today is obsolete in three years. In medicine and engineering, half of the knowledge learned today is obsolete in twenty-nine days. Obviously, a sense of powerlessness based on past perceptions rather than current realities could lead one to contribute to their own personal oppression. Human realities are in a constant state of change. Because one was powerless in the past does not mean one has to remain powerless. It may be easier for you to empower yourself today than it was in the past. Use your gift of self-awareness to determine the best place and time to begin your journey to personal empowerment. The empowered will seek out opportunities to fulfill the objectives of their agendas. The powerless will wait for entitlements from those they perceive to have power over them.

Returning to the elephant analogy, if the elephants were able to employ the human endowment of self-awareness, they would have the ability to assess their surroundings and their conditions, then use current, up-to-date information as a basis for their behavior. At that point, they might choose to pull themselves away from the line. With new information, they would have additional choices. What would the elephants do in an environment incapable of accommodating their massive figures roaming through the streets? Perhaps they would decide to go back and stand in line. In their new environment, they would certainly be faced with new decisions.

The simple truth is, we are always in a position of having to make choices. Our effectiveness as human beings is determined by the quality of the choices we make. And sometimes there are no good choices, but some options are better than others. The quality of our choices is probably best measured by how close they bring us to accomplishing goals we have set. We should consider how our choices relate to our life commitments. But first, we must clearly define our life commitments. Have we identified those things that are important to us, or are we conducting our lives based on someone else's agenda? Making responsible choices can be very difficult without having first developed personal objectives and ideals.

Personal agendas and self-perceptions have an enormous impact on how individuals behave during conflicts, and in many instances, determine whether conflicts develop in the first place. In order to resolve conflicts successfully or to avoid conflict altogether, I suggest that we look at situations and issues from a broader perspective than our own race or gender—or even our own politics, religion, sexual orientation, or any other categorization that can bias our views. We should put forth our humanism first—those qualities that are common to all of us as human beings.

In human interaction, most people respond to the ways in which we present ourselves. Some of us present our human elements and reserve focusing on race and gender. Others of us present ourselves from the personal perspectives of our race or gender, which automatically separates us from one another and subjugates our human elements. Most people will agree that our choice in this matter affects how others respond to us, despite the glaring obviousness of our race and gender.

Regarding the issue of race versus humanity, a White associate of mine once commented on the 1960s television series "I Spy" starring Bill Cosby and Robert Culp, a racially mixed detective team. This associate informed me that he did not think of Bill Cosby in terms of race.

He had chosen to view Cosby as a human being who was incidentally a Black male by birth. If my associate had employed an opposing paradigm, he would have seen Cosby as a Black male first and as a human being second. I was actually surprised because that was the first time I had heard a White person recognize a Black person as a human being first, rather than as a Black human being. Yet we are all human beings regardless of our race or gender. Recognizing the fundamental importance of this, I intend to expose the common thread between all people, which is our humanity.

We can harness more power with our humanity than with any personal perspective such as race or gender. The human endowments of self-awareness, freedom of choice, imagination, and independent will separate human beings from all other creatures. When we choose to use these gifts wisely, they will provide us with the foundation of our personal empowerment. When we make the choice to empower ourselves, we assume responsibility for our existence.

More so, if you choose the path of empowerment, you will be on your way to attaining your full potential as a human being. As you follow this path, you must choose how to present yourself. Will you present your gender and race, followed by your sense of humanity, or your sense of humanity, followed by your race and gender?

Some of us function in the larger community by commanding our birthrights as human beings, while others of us plead for or demand our civil rights. Those who fall into the latter group seek acceptance based on the personal perspectives of race, gender, sexual orientation, politics, and religion. In doing so, they sometimes meet with great resistance. It is my contention that commanding our birthrights as human beings places us in the position of empowerment that will bring respect and cooperation. Read along as we discuss the tactics of personal empowerment.

Chapter Eight

HUMAN ***BEHAVIOR***

AT THE UNIVERSITY OF ILLINOIS' annual Fire College Conference several years ago, I attended a presentation called, "What makes Johnny Run?" The instructor used a graphic model to illustrate the factors affecting the development and shaping of unique personalities that separate us from one another. I have included a copy of this model below. Note that there are three major elements that affect personality development and make each of us unique.

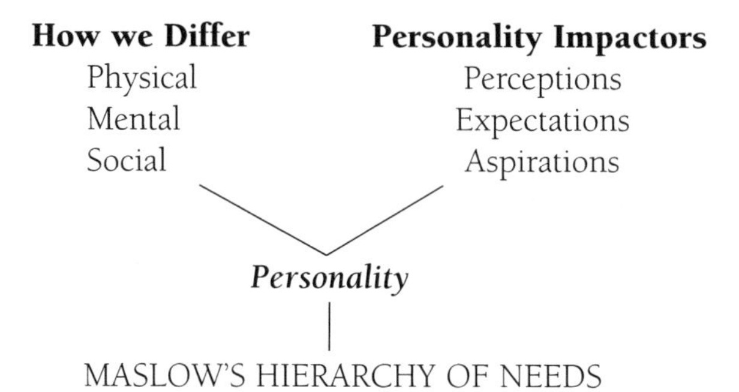

1. Biological: those needs necessary for life support.
2. Security: needs that protect and ensure continuity of the biological.
3. Social: needs of attention and belonging.
4. Self-esteem: needs recognition, status, and achievement.
5. Self-realization: needs opportunity to self actuate.

THE WAYS PEOPLE DIFFER

As the instructor explained (and the model shows), the factors that make people different from one another are some of the major influences in personality development. First, the instructor discussed how human beings obviously differ *physically;* some people are short, others are tall; some have a fair complexion, others have a dark complexion, and so on. He went on to explain how human beings also differ in their *mental abilities.* For example, some people have a facility with languages while others are exceptional with numbers. Then the instructor stated that human beings also differ *socially,* explaining that different backgrounds and past experiences produce different kinds of socialization. He said that acceptable behavior is learned rather than innate, and that what is considered "acceptable" varies according to each person's past experiences and social background.

The instructor not only used the model to identify how people differ physically, mentally, and socially, but also to identify similarities among people, or how we all have some shared characteristics and common ground in our humanity. And just as important, he emphasized that, although people do differ from one another in many ways, these differences are okay. "Different" is not bad or good; it is simply different.

We all have some shared characteristics and common ground in our humanity. For example, all human beings have dynamic personalities, that is, personalities that are capable of change. However, some people change more easily than others do. People with rigid personalities may regard changing themselves as a sign of weakness. They may be afraid that it will damage what they perceive to be their strength and stability.

Take a look at this example. One of the participants in a leadership workshop I conducted for fire officers told the group that his personality had been very rigid and unbending largely as a result of his 20 years of military training and service. He explained that he had great difficulty adjusting to civilian life after his discharge from the military. His lack of flexibility and inability to make the transition from a military to a non-military lifestyle left him feeling confused and frustrated. He sought help through therapy and eventually came to realize that it was okay to let go of some of his scrupulously maintained patterns of behavior and thinking. He could allow himself to yield to the needs of the present. In short, he could allow himself to adapt to his new way of life. He had discovered that the human personality is in fact designed to evolve and change.

As we turn corners and progress in our lives, it is quite natural to make changes within ourselves to promote our own well-being and happiness. The retired serviceman found that accepting change and learning to adapt alleviated the stress and inner conflicts he had been experiencing. His life was less traumatic and much more enjoyable; he became a healthier, happier person. So, if you or someone you know, find the word *change* distasteful or threatening, perhaps a more comfortable or acceptable word to use is *adapt*. As we go through our life's journey, we seek out opportunities to meet our personal objectives. We must adapt to whatever circumstances we encounter and accept and embrace our ability to change as the profound gift that it is.

PROCESSES THAT SHAPE AND RESHAPE THE PERSONALITY

As the model for human behavior illustrates, the second major contributor to personality development is the group of processes that fosters change in behavior and shapes personality; namely, our perceptions, expectations, and aspirations.

Perceptions

The way we make sense of ourselves, understand our roles in life and how they affect our behavior, and our understanding of the world shapes our thoughts and our concept of reality, which in turn influences our behavior. It follows then that misperceptions, because they are rooted in our sense of reality, may be difficult to alter. In order to test the validity of your perceptions, I recommend engaging in a discussion with someone from a different background who has had different experiences. An open exchange of views could lead to valuable new insights and clear up any number of misperceptions.

Expectations

Expectations influence changes in behavior. If people have low expectations, it is not hard to believe that they will act in accordance with those low expectations. Research and experience show that when expectations are raised, people often rise to the challenge and achieve higher goals.

Aspirations

Aspirations influence personality and behavior by functioning as sources of motivation. For example, what can minorities do about racism in America? If we aspire to overcome racism then our actions, behaviors, and energies will be directed toward that end. Our aspirations will act as a catalyst to influence our behavior, and to some extent, also shape our personalities.

FULFILLMENT OF NEEDS

The third major influence on the development of personality is the extent to which our human needs have been fulfilled. According to psychologist Abraham Maslow, there are five basic human needs that rank in order of importance:

1. **Biological.** These are the most important in Maslow's hierarchy because certain physical needs must be met to sustain life (*i.e.*, breathing or nourishment).

2. **Security.** This includes the need to feel protected, secure, and guarded against danger. We feel secure when we sense that our biological needs will continue to be met.

3. **Attention and belonging.** The need to receive attention and be accepted, to feel loved and cared for.

4. **Self-esteem.** The need for recognition and status.

5. **Self-realization.** The need to recognize our own individuality, to discover and appreciate our personal uniqueness.

There is a direct correlation between the satisfaction of these five needs and the development of our personalities and our resulting behavior. When our needs are met, we develop a positive personality and possess an inner happiness. On the other hand, when our needs are not met, we are frustrated and develop a negative personality; we possess an inner unhappiness. It is important to understand why our five basic needs are addressed in the order presented here. We cannot be aware of our security needs until our biological needs have been satisfied. We cannot be aware of our need for attention and belonging until our security needs have been met, and in turn, we are not aware of our need for self-esteem until our needs for attention and belonging have been met. And, finally, our need for self-realization does not surface until our needs for self-esteem have been met.

With this knowledge, it is not hard to understand why a person who—in order to get help in fulfilling a current need—may make some sort of deal with another person, but then fail to hold up his/her end of the bargain once that need is satisfied. When one need is satisfied, it is immediately replaced with another need, which can then rather easily take priority over fulfilling any commitments made to someone else to help satisfy a previous need. For example, a hungry person may promise someone that he will do certain chores in return for a meal. After he has eaten the meal, however, he may forget about the promise he made to do the chores. He may need to be reminded that his promise served as a condition for his receiving the meal. Since his need is satisfied, he is now responsible for making good on his promise. But, it is often difficult for some people to remember agreed-upon conditions when they are already engaged in trying to satisfy their next need. Their failure to follow through on their commitments will inevitably lead to conflicts with those to whom they made promises. This type of conflict has a simple solution, which is to remind the person of the commitment and hold them to their responsibility of fulfilling it.

Because conflict tends to be in lock step with change as a major constant in life, in the next chapter we will take an in-depth look at conflict resolution.

CHICAGO FIREFIGHTERS' STRIKE, 1980

The 1980 Chicago Firefighters' Strike was a major rage/guilt saga for me that had career-threatening or life-altering potential. I'll go back to the summer of 1979. As an instructor assigned to the training division working with active in-service companies, we trained members to perform three-man ladder raises. Historically, the Chicago Fire Department had been manning their hook and ladder companies with six or more people. Standard operating procedures required four to six members to raise most of the ladders carried on those apparatus. Now with a manpower shortage, we were teaching a three-man ladder raise to get the job done with fewer people. Many hook and ladder companies were running with an officer and three firefighters, and in order to carry out their rescue duties they had to learn new methods to get their ladders raised to victims on upper floors.

In November 1979, a promotional order came out. This order promoted a large number of lieutenants and engineers. The city also hired additional firefighters at this time to relieve the severe manpower shortage. The department had been below budgeted strength for approximately two years. This was a serious concern for the union. As a safety issue, it is dangerous for four people to do the work of six when the workload remains the same. The mayor, Jane Byrne, was also perceived to be less than truthful in her dealings with the public. It seemed that she would say one thing and do another. That practice resulted in a strong sense of distrust.

For years, while Richard J. Daley was mayor, the firefighters worked under a verbal agreement sealed with a handshake. Frank Muscare later ran, and was elected president of the Chicago Firefighters Union, on a platform to get his members a written contract. Many firefighters were uncomfortable doing business based on a verbal agreement with a person they perceived to be dishonest. The firefighters believed that Muscare would be able to force the city to the bargaining table and secure the first written work agreement. After several failed attempts to get the city to return to the bargaining table and continue negotiations, the union called for and received a strike authorization vote from its membership. Contrary to popular belief, the major reason for the success of the vote was the manpower shortage. There was still a shortage of manpower even after the new hires and promotions in the fall of 1979. The lack of trust between the union and the city was a contributing factor.

During the predawn hours of February 14, 1980, I received a phone call from my good friend John Eversole informing me that Muscare and the union had called a strike, and that firefighters were walking out of the firehouses. He directed me to report to my district headquarters. I tried unsuccessfully to get back to sleep, thinking that it was all a bad dream. If not, it would all be over by noon. I felt that this could not be the real thing, because most firefighters liked their jobs. Yes, we wanted a written contract, but to actually walk out of the firehouse—no way, I thought. After getting out of bed, I began to move slowly and deliberately, wondering how best to handle this if it was the real deal. A few hours later I arrived at district headquarters, Engine 95's quarters. Sure enough, there was a picket line blocking the entrance. I decided that I was not going to cross that line.

There were several reasons for that decision besides believing in the labor movement and believing that I was promoted to lieutenant because of union pressure on the city to hire and promote. I also thought that it was reasonable for the city to give us a written labor agreement. It was not about money. We wanted the current verbal agreement put into writing.

The tough part was going against the oath I had taken to serve and protect the public. It was a very hard decision for me to make, and I experienced a sense of guilt afterwards. I rationalized that the city had a responsibility to be reasonable with its employees whether sworn or not. Refusing to provide us with a written labor agreement was unreasonable. I thought that certainly by noon a level head would surface and the city and union representatives would meet and resolve this strike.

At Engine 95's quarters, I met with my good friend Stanley Span and went to a near-by coffee shop to have coffee and lots of conversation. We agreed that this would be a short-lived situation. The fact is, we were wrong. For the next few days I met with a group of Black firefighters and supervisors. We struggled to make some sense out of the situation. There we were on strike with White firefighters, supporting an organization that, historically, had been less than accommodating to its minority members' concerns. Captains Ado Warren, Carl Garnes, Francisco DeLaCerna, and firefighters Landis McAlpin, Roy Carroll,

James Winbush, Leslie Noy, and myself assumed leadership roles in this ad hoc group of Black firefighters. The thinking among us was that it would be beneficial for us to side with the union in their effort to secure a written contract with a mayor whom they could not trust. With the union shaking things up, at least the status quo would be disturbed and perhaps we could make some progress in our dealings with both sides—each of which simply ignored or, at best tolerated, our presence. Further, it was thought that our presence on the picket lines would give us credibility with the union after the strike ended.

One of the things we came to realize was that we were on strike and the citizens had no idea what the reasons were. The decision was made to go to the public and keep them updated as to the issues and the progress, if any. Having abandoned our commitment to serve and protect them, the very least we could do would was keep them informed. There were both supporters and opponents in the public arena, but mostly there was confusion in the community. Because the minority communities were the primary users of our services, we felt that we had more responsibility and a deeper commitment to reduce this interruption of services to them.

On Saturday,February 16, 1980, with this in mind,Landis McAlpin, an ordained minister, Francisco DeLaCerna, and I set up a meeting with Rev. Jesse Jackson at Operation PUSH. We updated him on the issues and our concerns. He was quick to offer his support saying that he understood our dilemma and would have his people get back with us in a few days. The Rev. Willie Barrow and I had an opportunity to talk about these issues after church the next day. She said that our situation was not unusual for minority members of large organizations and that we should keep the faith in our effort.

The next day (Monday), I received a phone call from Rev. Barrow asking if our group could meet Rev. Jackson at his Operation PUSH offices. As fate would have it, there was a meeting at my home when the call came in. Those of us present began a call-around campaign to get as many people as possible to attend the meeting with Rev. Jackson. Upon our arrival at the meeting, we were interviewed by some of Rev. Jackson's staff. We provided them with as much background information as possible. There was a general agreement that neither the city nor the union had been appropriately responsive to minority concerns. As was demonstrated by the absence of minority representation in the union's and city's leadership. Both had a long history of insensitivity to minority issues. We were in a position of having to choose the lesser of the evils.

Because of this circumstance, we could not blame the minority members who refused to honor the strike. It came down to members having to decide from which position they wanted to address their interests. There was rage and guilt on both sides. Rage because one side would not honor the legitimate requests of the other, and guilt because the public's trust was not being honored.

A short while later, Rev. Jackson informed the group that he had been in touch with Mayor Byrne's office and had set an appointment to meet with her the next day, Tuesday, February 19. Representatives of the union, the fire commissioner's staff, Fire Commissioner Richard Albrecht, and members of the mayor's staff were present along with Rev. Jackson and our group. The mayor expressed her opposition to any further discussions with the union because in prior meetings the union had changed the agenda as the items on the agreed upon agenda were resolved. She said it was as though the union wanted a strike more than a resolution of the issues. She had decided that her staff would not participate in this kind of behavior. Rev. Jackson reminded her that she had a moral responsibility to talk and work toward a resolution of this crisis. He assured her that the type of behavior she accused the union of would not be tolerated. She then informed us that the strike was 97 percent effective and that the fire department's commissioner had failed her by underestimating the union's support. Rev. Jackson then asked the members of the union and the fire commissioner's staff to meet him at a community meeting that he was calling for that evening at the headquarters of Operation Push.

At that meeting, Operation Push had members of its labor coalition in attendance. The Firefighters' Union officials were put on the spot as they were asked about the minority representation on their executive board. The community and the labor coalition leaders were shocked and disappointed upon learning that there were no elected minorities in the firefighters' organization. The representatives of the fire department were informed that the lack of minorities in policy-making positions within the fire department was a disgrace and would not be tolerated after the strike was over. That meeting and those discussions resulted in the affirmative action clause being put into the contract. Mr.

Charles Hayes and Mrs. Addie Wyatt, who represented Commercial Food Workers of America as president and vice president respectively, led the Operation PUSH labor coalition. Mr. Hayes, who was later elected to the United States House of Representatives, called for a massive Chicagoland labor rally to support the striking firefighters and their families. The rally was held at the Chicago Amphitheater. There were several thousand people there representing the Chicago Federation of Labor and it's member unions.

During the following weeks, the firefighters' international and local unions held several rallies and strategy meetings. Several attempts were made to get back to the bargaining table. On Saturday, March 1, I was asked to be available for a radio program hosted by attorney E. Duke McNeil for the purpose of answering questions from the public relative to the strike. During that program, William Kugelman, a leader of the union's negotiating team, called in and announced that the union's representatives would be at the Bismarck Hotel, across the street from city hall, the next week prepared to work with the city toward a resolution of the strike.

As minority strikers, we found that we had access to both camps because, for some reason, neither side felt threatened by our presence. I was given the assignment to meet with the fire commissioner's staff in an effort to identify their issues. Much to my surprise, there was more conflict among the staff regarding a major power struggle within the group. They seemed to be more concerned with who would be in charge when the strike was over than they were with labor problems. Their conflict was caused by the fact that the deputy fire commissioners on the fire commissioner's staff had been promoted after passing a civil service examination. With civil service certification, they could not be easily removed from their positions. There was some resentment when Richard Albrecht was appointed fire commissioner. Before his appointment, he was a captain who had served as an administrative assistant to former Fire Commissioner Robert J. Quinn. Each of his deputy commissioners was a highly respected fireground commander with many years of experience. Prior to Commissioner Albrecht's appointment, fire commissioners were selected from certified chief officers. This break with tradition by the first female mayor of the city did not help calm the working environment in a department already in turmoil.

It soon became clear to us that we had a situation on our hands where both sides had gotten themselves into a bind and did not know how to work their way out of it. At one of our meetings, we decided that we had to get back to work and protect the community as soon as possible without embarrassing the union. At our next meeting, the union's attorney, J. Dale Berry, informed us that he had advised the union against going on strike. Under further questioning, he said that Michael Lass, the representative from the International Association of Firefighters, had convinced the union's executive board to strike. For clarity, Captain DeLaCerna asked him: "Do you mean to tell us that we are on strike against your advice, as our lawyer?" The answer was, "that's correct". Several members of the group left the meeting and went back to work.

At that point, the strike was not about the many legitimate labor concerns that would be addressed successfully later, but rather, it was about a series of power struggles and individual agendas. On the city's side, there was a fire commissioner in a power struggle with his command staff. On the union's side, there was a power struggle between their international representative and their attorney. This conflict was resolved when Mr. Lass left town for reasons unknown to me. Central to all of this was the power struggle between the union and the mayor.

On Thursday, March 6, Rev. Jackson led an interracial group of striking firefighters along State Street in downtown Chicago. During the march, a White firefighter said to me, "I never thought that I would see the day that I would be marching behind Jesse Jackson singing We Shall Overcome." We Shall Overcome is a standard song for civil rights campaigns.

After that march and the rally that followed, Rev. Jackson called for and set up a series of meetings to begin that evening at the Bismarck Hotel. The purpose of the meetings was to resume negotiations between the city and the union. With the public watching, it was important for both parties to be perceived as eager to get the strike resolved. The meeting was structured in a manner that allowed each group to meet in separate rooms. These meetings lasted well into the night and continued on throughout the next day. Rev. Jackson moved from one caucus room to the other keeping each one focused on the labor issues at hand, and delivering communications from one group to the other. He was effective at clarifying the concerns of each side.

Operation PUSH had assigned minority community leaders to each room as monitors to observe the discussion and to assist as needed.

There was a sense of urgency in the air, and fortunately, there had not been any major emergency incidents in the city during the strike. We all knew this situation could not last much longer. On Friday, March 7, Rev. Jackson called in a labor attorney and a secretary to reduce the verbiage down to a written statement. The statement became the first labor contract between Chicago Firefighters Local #2 and the city of Chicago. The contract was ratified at a quickly called mass meeting at McCormick Place Inn that night, and the firefighters returned to work on Saturday, March 8.

The fire commissioner had received less than accurate information from his staff. He in turn, passed this inaccurate information on to the mayor who then made decisions based on the misinformation. Would she have made the same decisions had she been informed that 97 percent of the firefighters were willing to walk off their jobs? She implied she would not have, and said that the fire commissioner, as the administrator of the fire department, had the responsibility for keeping the firefighters on their jobs. Did his staff officers provide him with full disclosure of the information available to them? We don't know what they knew or when they knew it. On the other side, if the union had followed the recommendation of their attorney, they would have made a different decision. There was a conflict between the representative from the international union and the local union's attorney. The international union's representative, Michael Lass, recommended the strike over the objection of J. Dale Berry, the local union's attorney.

Public Pressure forced both sides to put aside their internal conflicts and focus on the labor issues. To serve the needs of a neglected community without doing unnecessary damage to either side, the experienced labor negotiators and monitors from the Operation PUSH labor coalition guided the proceedings to a successful resolution.

After the contract was ratified and the firefighters returned to work, in addition to the resentment that one could expect, there was a strong presence of the rage/guilt syndrome. The members who worked during the strike experienced a sense of rage because the strikers were not fired, and the members who were on strike experienced a sense of guilt because they had violated their oath to serve and protect the community. As the benefits from the strike settlement were realized, some of the members who had worked during the strike were so guilt stricken that they didn't feel comfortable accepting the benefits. Several of them sought out and received professional help in order to address their feelings of guilt.

Chief William Foran made, what I considered, a profound statement that addressed the rage/guilt syndrome. He said, "If you did what you believed in during the strike, you have nothing to be ashamed of." He went on to suggest that those who chose to walk out or to work because they were following others and not the dictates of their own personal values would have more problems adjusting after the strike. When we behave in concert with our internal values, we tend to be more comfortable with the results. The effects of the rage/guilt syndrome are reduced when we conduct our affairs from our heart-of-hearts.

In January 1985, I was called into a meeting with the new fire commissioner, Louis Galante, and his deputy commissioner of Fire Suppression and Rescue, Ronald Maloney. Commissioner Galante asked if I could work with the director of training, George Malik. I told him that George and I had worked together before successfully and that I could not see any reason we could not work together again. Much to my surprise, he said that he needed to get some effective training done and he thought that I was the person to help him get it done as the assistant director of training. I told him that I was very honored by his offer, but I thought that my personality might be too crass for the politicians that I would be expected to work with. He assured me that he would take care of the politicians if I would take care of the training. Commissioner Maloney added that, "If you change, you will be removed from the position." He continued, saying that my direct no-nonsense manner was needed for the troubled state that the department was in at that time. I told him that I would have to think about the offer, which would require that I resign my career service status to accept the exempt rank position on the fire commissioner's staff. After giving it some thought, I accepted the offer with the understanding that I would be reinstated to my career service rank of lieutenant should we not get the desired results. This was only the second time in the department's history that a lieutenant was given an appointment to an exempt rank position. The first time this occurred was approximately two years earlier when Lieutenant Michael

Shanahan was appointed assistant director of the repair shops. Such an appointment means that the lieutenant promoted to that position skips over the ranks of captain and battalion chief.

I accepted the position believing that the basis of the hostile atmosphere that existed throughout the department was created by a series of misperceptions made during the strike, and sustained afterwards. My first concern was to gather a group of instructors who could commit to providing effective training. I was successful in retaining most of the experienced instructors already assigned to training. I called a meeting of all the instructors who were on the training staff when I took over. I informed them that I understood that they had not planned to work for me when they accepted their positions and that I would understand if they were uncomfortable working for me. I suggested that each of them go home that night and decide whether or not they wanted to work for me. If they decided that they did not want to make a commitment, I would assist them in getting the assignment of their choice. I was very pleased that most of the staff agreed to stay to assist in providing an effective training program for the department.

A request for instructors to fill the staff vacancies was sent out. We were very successful in attracting experienced line officers. I conducted a two-day, train-the-trainer seminar for the entire training staff so that we would all be working toward the same objectives in a similar manner. Commissioner Maloney informed us that we should prepare our facilities and programs to train 100 new recruits, 20 of whom would be females. The females would receive physical training before entering the academy. That meant that only minimal adjustments in physical performance standards, if any, would be necessary to accommodate them. The next project the commissioner wanted us to prepare for was cardiopulmonary resuscitation (CPR) training for the 5,200 members of the department. This would be difficult in the hostile emotional environment that existed at that time.

The decision was made that I would conduct a conflict resolution class in an effort to reduce the tensions. The major problem was getting such a large number of people through the program in a reasonable timeframe. We scheduled 12 working companies with five members each to attend each 3-hour training session. By conducting two sessions a day for eight weeks, approximately 4,800 members were trained. The Human Behavior, Behavior-Attitude-Value, and Conflict Resolution models described in this book were used in these classes. In addition, I created work groups consisting of members from different companies. It was critical that the groups be mixed in a manner that would have as many strangers working together as possible. That way we had strikers and non-strikers working together on common problems. There were 10 groups of 6 members each. Each group was given a station personnel problem to resolve and they were to then present a report to the reassembled class. The result was a 50 percent reduction in union grievances. Before the beginning of the classes, there was an average of 100 grievances per month filed against the city by the union. Eight weeks later at the end of the classes, that average was below 50 per month. I am told that this lower monthly average has continued since that class.

The Chicago Firefighters' Strike and the bitterness that followed for five years after it, was based on negative attitudes, distrust, misperceptions, and incomplete/inaccurate information. As these matters were addressed appropriately, the tensions were reduced and cooperation increased, not only in the workplace, but also at home among families with multiple members on the department. It is very important that two-way communication be maintained at all times. Participants must also put thinking before emoting by identifying the issues and minimizing the attitudes. In so doing, we ought to be able to agree to disagree without being disagreeable. Very often, time is saved and hard feelings are spared when each party identifies their issues up front. "What do you want to come of this discussion?" is a question that each party can ask of the other. The answers could surprise you. Many times the answers indicate that there is something that each party can give up without suffering a major loss, or the answers can provide a foundation for productive negotiations.

Satisfaction of our "Hierarchy of Needs" as described by Dr. Maslow leads to positive self-perceptions, positive expectations, and positive aspirations. The positive effects on these personality impactors contribute to a positive personality, increased self-worth, and positive behaviors in people.

Most often in the work place the "Biological" and "Security" needs have been met. That leaves the needs of "Attention and Belonging", "Self-esteem", and "Self-realization" to be addressed. The following is an example of how I addressed the self-esteem need in the work place. As the district chief it was my responsibility to maintain the morale of the members assigned to the district. Utilizing the concepts detailed in this model, each year on their birthday the members received a birthday letter from me at their home. In addition to wishing them a happy birthday and many happy returns, I thanked them for their help in providing the best possible fire protection for the citizens that lived and worked in our district. This extra bit of positive recognition resulted in increased personal self-esteem and increased energy directed toward their work. Our district became known as a good place to work.

The real test for this model is personal. When your needs (as described here) are met, do you develop a positive personality and inner happiness? If your answer is "yes" you have reason to believe in its effectiveness. You will be able to apply this concept with confidence. Should your answer be "maybe" or "sometimes" you will have to develop confidence in the concept by identifying opportunities to use it. When faced with a trouble employee a good first step is to demonstrate concern by providing support and positive recognition if appropriate—even discipline can be applied in a positive manner.

Chapter Nine

CONFLICT RESOLUTION

CONFLICT OCCURS when the goals of one person are incompatible with the goals of another. Clashes of interest and ideas are inevitable, so the question is not how to avoid conflict, but rather how to resolve conflict. No matter what the outcome of the dispute is, the approach that each side uses in resolving the struggle will have a lasting impact on the relationship between the two parties. When one or both of the parties enters the conflict with defensiveness, distrust, hostility, or dishonesty, tensions between them escalate and the conflict is painfully prolonged. Then when a resolution is eventually reached, the relationship between the two parties has been damaged. When conflicting parties take the opposite approach, however, their relationship can be remarkably improved. Discussing the issues of contention in a very open and honest way can assist in resolving problems. Putting forth concerns in a confident, accurate, and non-combative manner helps to keep tensions at a minimum and to accelerate the process of conflict resolution. The end result is often that the two parties have an enhanced relationship with greater respect and concern for one another.

Some authorities, the National Fire Academy among them, report that conflicts involve 20 percent substance and issues and 80 percent attitude and emotion. What must be understood is this:

- Substance can be defined. The facts about an issue can be stated and addressed in objective, and relative terms.
- Attitudes and emotion cannot be measured or explained in definitive, objective terms because they reflect feelings.

In conflicts and in life, we are not responsible for justifying our feelings, but we are accountable for our behavior. We may disagree, but we do not have to be disagreeable. Allowing ourselves to be rude or combative does not achieve our goals; in fact, such behavior weakens your position and ultimately gives our opponent a measure of control over us. If an open mind and the most earnest efforts fail to yield a satisfactory resolution to a conflict, then both parties must simply agree to disagree.

When a conflict arises, you have to decide when and how to respond to the other party, and then put forth your position. These initial decisions will shape the character and outcome of the conflict. You may decide to respond at the time of your choice, and in your own manner, according to the objectives of your personal agenda.

When engaged in a conflict, there are two important things to remember: (1) protect your interest, and (2) avoid causing unnecessary damage to your relationship with the other person. In fact, if it is possible, you should even try to improve the relationship.

To protect your own interests, you need to make it clear to the other person that you will not assent to any resolution that you feel is not beneficial or fair to you. You must preserve your own well-being. You have to be firm and assertive in advocating your rights and in doing what is best for you. But, at the same time, you must try to keep the discussion focused on the issues at hand without letting your emotions get in the way of achieving your goal of resolving the conflict. In a highly emotional state, you or the other person can easily cause serious damage to your relationship. Rather than focusing on the substance of the conflict, one person may bring up unrelated sensitive issues, make accusatory statements, or resort to unfair tactics that will surely alienate the other person and aggravate what may already be a precarious relationship. At the very least, lack of concern for the other person's interest results in a weakened relationship with that person. This weakened condition may, in turn, trigger new conflicts in which the prevailing attitude of both parties may be, "Well, he doesn't care about me, so why should I care about him?" If, on the other hand, you can respond by demonstrating sincere concern for the other person's interest, and try to find ways to meet the objectives of both sides, you may be able to avoid future conflicts with that person and actually build a stronger and healthier relationship.

TYPES OF CONFLICT

Conflicts that develop in our personal life or our organizational life usually fall into one of the following categories:

- **Role conflict.** These conflicts stem from differences of opinion that develop between superiors and subordinates or between parents and children about "on-the-job" issues, such as responsibilities, rights and privileges, priorities, responsiveness, attitude, and conduct.
- **Interpersonal conflicts.** Sometimes called personality conflicts, these conflicts generally occur when individuals have different motives or are struggling for increased power. They often occur between people at similar stations in life or between siblings.
- **Intra-group conflict.** These types of conflicts occur when subgroups disagree with one another concerning operational issues or issues relating to which subgroup will have the most influence in the organization. In-laws or other relatives sometimes experience these kinds of conflicts.
- **Inter-group conflict.** These are conflicts that can develop when members of a group have different duties and/or responsibilities. For example, staff officers are concerned with policy development and planning. Field officers are concerned with implementing policies. Field officers want as much information as possible as soon as possible. It would be irresponsible, however, for the staff officers to pass information to the field prematurely. Staff should have assurances that a given project is viable before informing the task providers of the project. Should the project be determined undoable, the organization's leadership could be perceived as unstable.

Differences between a labor leader and management personnel are probably the most common source of inter-group conflicts within organizations. Labor leaders have different areas of concern than managers. The primary responsibilities of managers are administering departmental policies and approved practices,; whereas, the primary responsibility of labor leaders is monitoring and protecting the welfare of their members. So, in the course of carrying out their jobs, people from both sides often find that their goals are not compatible, which then results in conflict.

Understanding conflict

In order to deal successfully with a conflict, we must accurately diagnose the conflict in question. This involves three major steps:

1. Determining the current stage of the conflict.
2. Discerning the nature of the conflict.
3. Identifying the factors that underlie the conflict.

Stages of conflict

From inception to resolution, most conflicts occur in four stages: frustration, analysis, behavior, and outcome. Let's use an example to illustrate the progression of these stages:

A manager wants to make some changes in policies and procedures. In order for the changes to be successfully implemented, she will need the support and cooperation of subordinates. A conflict arises when the new regulations are slow to be adopted.

Progression of conflict stages:

1. **Frustration.** Frustration occurs when someone or something is blocking an individual or a group from obtaining a desired goal. Although the source of the frustration may not be well-defined at this stage, an individual or group may nevertheless feel thwarted because the desired result has not been achieved. In our example, the manager becomes frustrated when satisfactory progress is not

made in adopting the new policies. Her frustration is compounded when she cannot pinpoint the reason why the changes haven't been fully implemented. The workers are also frustrated because their input was not invited or appreciated when offered.

2. **Analysis.** At this stage of a conflict, the nature and scope of the problem causing the frustration can be clarified. In our example, the manager is able to answer the question, "What is the real problem?" Perhaps the new policies and procedures are being resisted because some of the workers resent the fact that the changes were decided on without their input.

3. **Behavior.** This is the stage during which some action is taken to deal with the cause of the conflict now that it has been conceptualized. The manager, for instance, might set up a meeting about the new policies and procedures with all the parties concerned. During this phase of a conflict, a neutral moderator can be very useful.

4. **Outcome.** The outcome stage is of course the stage during which resolution occurs. After action has been taken and decisions are made, the results can then be evaluated. In our example, the manager can encourage discussion, make accommodations, and then effectively assess the quality of the resolution by asking the following questions:

    - Are the results satisfactory to both sides?
    - Have the interests of both parties been considered and protected?
    - How has the relationship between the two parties changed? Is it weaker or stronger? Is it more reticent and distrustful, or more open and honest?

Sources of conflict

Understanding the evolution of a conflict enables us to rationally approach conflicts. Realizing the need to clearly understand the nature and scope of the issues helps to head off the greater conflicts that often arise when action is taken at the frustration stage before the problem has been clearly defined.

In addition to understanding the four stages of conflict, it is also helpful to recognize the various types sources of conflicts. Typically, conflicts fall into one or more of the following categories:

- **Facts.** We may disagree because we have different information concerning the problem. Do we all have the same information? Is the information accurate? Is the information complete?

- **Methods.** Disagreements may arise over which procedures or methods are best for achieving a given goal.

- **Goals.** There may be disagreements over what the final results should be.

- **Values.** Disagreements regarding ethics can arise. Moral considerations, like whether the end justifies the means, are sometimes the source of conflict.

Understanding the nature of a particular conflict is of vital importance to us. Resolving a conflict over facts, methods, or goals will be easier than resolving a conflict over values. Therefore, this step prepares us in an important way. We can approach the potential resolution of a conflict with a clearer perspective on how difficult that process is likely to be.

There are often various factors that underlie conflict. They include one or more of the following:

- **Information.** Sometimes conflicting parties draw different conclusions based on varying information. The ancient fable of the six blind men feeling different parts of an elephant and each concluding that he is experiencing some object other than an elephant is reflective of this type of underlying factor. In other words, parties not having access to the same information are one factor of conflict. All parties should have the same accurate, complete information that is appropriate for their level of participation.

- **Perception.** Each person filters new information through a unique set of past experiences. These filters affect and alter the way different people interpret the same information. In other words, varying perceptions regarding the same information is another contributing factor to conflict.

- **Role.** Often with a change in role, there is a change in concerns. As employees advance into management positions, for example, their positions in conflicts may change with their role change. Roles, then, like varying information and different perceptions of the same information, can also contribute to the development of conflict.

Identifying probable underlying factors in a conflict can greatly improve your chances of selecting a workable resolution strategy. It is important to remember that conflicts have two aspects: (1) logical, having to do with issues or facts, and (2) emotional, having to do with attitudes and feelings. The effective negotiator pays attention to both. When you reflect upon your past conflicts, you may recall experiencing a number of these feelings such as the need for power or control, a low sense of self-esteem, defensiveness, or anger. By honestly recognizing these issues, and openly addressing them before they escalate, it is sometimes possible to defuse a potentially damaging conflict. Now, let's look at the way we communicate, and, subsequently, how communication incites or defuses conflicts.

Chapter Ten

COMMUNICATIONS *MODEL*

HAVE YOU EVER been engaged in a discussion with someone, and without really knowing why, felt compelled to defend yourself as if you were being personally attacked or accused of something? These defensive impulses are often stimulated not so much by what someone else says, but rather by the terminology that he or she uses. For example, the words *you*, *your*, *they*, and *them* are terms of reference that can trigger defense mechanisms and promote a feeling of isolation, separation, and division. They are exclusive pronouns. In contrast, the inclusive pronouns *we*, *our*, and *us* are terms of reference that tend to foster a willingness to work together and to create a sense of unity and harmony.

Effective communications

Being aware of the impact that our word choices can have on a listener improves our ability to communicate effectively and to avoid unnecessary misunderstandings. We are most likely to achieve the outcome we desire when we take a positive and collaborative approach to a discussion and when we use terms that are compatible with our intended meaning.

One way to sharpen your communication skills is to listen to yourself from the other person's perspective. This simple exercise can be very useful in helping you recognize and identify any negative (and usually self-defeating) communication habits that you may have developed.

Mentally switching roles can help us understand how some words and behaviors, despite our good intention, might be perceived as self-serving, manipulative, offensive, or demeaning. For example, while we may be trying to win a listener's favor, if we hear ourselves repeatedly using egocentric terms like *I, me,* and *mine,* the listener will begin to feel belittled or alienated. Once we grasp the power that our words and behaviors have, we can sidestep misinterpretation and successfully communicate our message.

There are also things we can do when we are on the receiving end of what appears to be manipulative or offensive words and actions. First, we can remind ourselves that often when people regularly boast of their achievements, consistently make excuses for their failures, or take some other negative approach to communicating, *they are actually seeking approval.* Their methods may seem offensive, but their intentions are not. Knowing that their offensiveness probably stems from their insecurity, we are more likely to be annoyed than angry. Sometimes with a simple nod or statement of approval from us, their language will reflect increased confidence, their approach will become more positive, and their terminology will become more inclusive.

Most of us have difficulty thinking and emoting at the same time. For those times when we find ourselves in heated discussions, an appropriately placed question beginning with *Who, What, When, Where,* or *Why* will tend to reduce the emotion and increase thinking. These terms will lead toward decisions based on information instead of feelings. When we choose to place the "T" (thinking) before the "E" (emotion) we will make the more effective decision. The basis for our decision-making should be the stated goals of our personal agenda, and the process or behavior we choose should also be in concert with those goals.

At one time or another, we are all on either the sending or the receiving end of an important message being communicated. No matter which end we are on, we must do what we can to reduce the possibility of unnecessary misunderstandings or hostilities. We can function in a problem-solving mode by using terms like "let's do..." and "together we can..." Such words and phrases can change attitudes and draw people together. If we can learn to really listen to ourselves and choose the right words, we will have found the key to effective, successful communication.

Chapter Eleven

BEHAVIOR~ATTITUDE~VALUE **MODEL**

MANY AUTHORITY FIGURES accept responsibility not only for the *well-being* of those in their charge, but for their behavior as well. In an effort to ensure that the behavior of their subordinates is acceptable under the standards set by the company or society as a whole, supervisors often spend a great deal of time trying to change the attitudes of others. Their logic is that if they can change poor or unacceptable attitudes, proper and acceptable behavior will follow. The flaw in this logic however is that those in authority are responsible for evaluating and regulating *behavior,* not *attitudes.* Behavior can be addressed and documented more directly and definitively than attitude. It can be described in clear, concise, objective terms. Descriptions of attitude, on the other hand, tend to require lengthy, drawn-out, subjective terminology.

Organizations can write regulations based on acceptable or unacceptable behavior to clearly describe policy. This ensures that supervisors can effectively administrate and enforce the organization's policies in clear, concise, objective terms. This enables business to be conducted in an orderly atmosphere. On the other hand, regulations based on acceptable or unacceptable attitudes tend to be confusing and subject to the attitudes of both the writer and the reader. Policies developed from attitudes contribute disorder within organizations. As we discussed in the section on conflict resolution, attitudes and emotions cannot be measured or explained in definitive, objective terms; they reflect feelings. The result is that discussions about attitudes tend to be much more contentious and far less productive than discussions about behavior. In cases where attitude is seen as a problem, the common good might be best served if the person in authority first identified examples of the other person's undesirable or unacceptable behavior and then followed up by asking that person to talk about his or her situation and attitude.

These types of discussion provide an opportunity for the effective use of questions beginning with, *Who, What, When, Where,* and *Why.* As we mentioned in the Communications chapter, developing answers for these questions not only tends to reduce tensions, but also tends to de-emphasize the impact of one's attitude during a discussion. There can be a multitude of reasons contributing to someone's attitude that have nothing to do with the supervisor, so it is always best to give the person a chance to be heard before independently developing an uninformed conclusion.

A case in point is a firefighter assigned to a company in my district. He asked to meet with me and told me that his wife was having an affair and neglecting his children while he was at the firehouse. He wanted help in preventing his personal problems from becoming work problems. It was vital to him that I understand that the fire department was important to him. He explained that when he seemed out of sorts or inattentive, the problem was not lack of interest in his work, but that he was having a tough time with his personal life. With this knowledge, I agreed to support him when and if he ran into difficulties on the job. He also followed my suggestion and talked to the professionals in the employee assistance program. As this example shows, a person's attitude may very well be a reflection of issues unrelated to his or her supervisor or job. In many cases, attitude is actually a call for help. And when it is, the person needs assistance instead of a reprimand, compassion instead of punishment. In the real world, people are responsible for their behavior. The evaluation of behavior can be done with objectivity; the evaluation of attitude, however, is far more subjective. While unacceptable behavior must be addressed appropriately, no one has the right to judge the worthiness of another as a human being.

Truth resides in every human heart.
One has to search for it there, and be guided by the truth
as one sees it. But no one has the right to coerce others
to act according to his own view of truth.

~Mahatma Gandhi~

Chapter Twelve

WINNERS ~VS~ LOSERS

WHAT ARE WE personally responsible for? This was the question that came up during a leadership seminar that I conducted for a group of fire service supervisors. After some discussion, we agreed that each of us is responsible for our own happiness and well-being, for protecting our own interest without doing unnecessary harm to others, and for behaving in accord with our personal values. How is personal responsibility related to success and failure? What distinguishes a winner from a loser?

I think the difference between success and failure lies in the level of commitment brought to developing and carrying out a personal agenda. As we discussed in chapter 2, a personal agenda identifies the principles and personal values that govern our behavior and serves as a guide for establishing and fulfilling our personal goals. Meeting these personal objectives sometimes involves self-discipline and tough choices, and how we respond when faced with those tough choices is what separates a winner from a loser.

The student who spends his time in the student lounge playing video games instead of in the library doing his course work is an example of a loser. Near the end of the semester, he talks to the professor to see if there is anything he can do to get a passing grade. Even if he does manage to pass, he has still lost out on the full learning experience that his tuition had entitled him to. Winners on the other hand, complete their course work before participating in leisure-time activities. Winners get more value out of their tuition dollar and more enjoyment from their leisure-time activities because they can use their leisure time single-mindedly without the additional burden of having to work on a scam for the professor. These leisure activities then become recreational in nature because they actually re-create energy for winners.

My definition of a loser is a person who behaves contrary to his or her own best interests. Like the student who chooses to spend time playing video games instead of studying, losers use their resources and opportunities to do what they want to do before doing what they need to do. Winners though, put their responsibilities first, their desires second. They use their human endowments and their self-discipline to focus on their commitments and to achieve the goals set out in their personal agenda. And, not surprisingly, they are successful more than 90 percent of the time.

There are no born losers or winners, but success often seems out of reach to people who behave irresponsibly without the motivational support or self-guidance of a personal agenda. These people tend to rationalize their failures as someone else's fault or as a result of circumstances beyond their control. People who behave responsibly, on the other hand, don't offer excuses or blame others for their failures. They are also more likely to keep trying to reach their goals until they succeed in doing so.

Moreover, winning versus losing is about accepting responsibility for our behavior and about following through on our commitments. It all comes down to making choices. Do you want to be a winner or a loser? The choice is yours.

Chapter Thirteen

EFFECTIVE *LEADERSHIP*

MOTIVATING WORKERS

SUPERVISORS ARE RESPONSIBLE for motivating the people on their staff, and in that pursuit the most effective leaders generally follow five steps. Because effective parenting skills overlap effective leadership skills in many areas, these five steps can be employed equally as well in the home setting to motivate children.

1. **Tell people what is expected of them.** Sometimes supervisors fail to communicate (either orally or in writing) their expectations exactly as they expect their subordinates to carry them out. Supervisors may simply assume that their expectations are obvious or "understood", but such assumptions are unfair and even counterproductive. If you expect a particular task to be performed or specific goals to be accomplished, then you should enumerate them. Be clear, and be specific. And just as important, be sure to tell workers why these tasks need to be done. If the reasons for the tasks are known and accepted, then methods for completing the tasks can be agreed upon.

2. **Make sure that people know how to complete their tasks.** Supervisors must recognize training as one of the most important elements of their jobs. Without appropriate skills, knowledge, and support, workers will be unable to achieve the objectives that are set for them. *Giving an employee responsibility without instruction is an unfair burden and a risky proposition.* It is always important to make the effort to get involved and to teach employees how to carry out their work successfully. Providing strong direction and guidance in the beginning usually leads to confident and independent performers in the end.

The value and mutual benefits of supervisors taking an active role in training subordinates was clearly impressed upon me after a situation unfolded in the fire department involving a company officer and a firefighter. The officer told his chief that one of his firefighters would not do his job correctly and asked for the chief's advice. The chief asked the officer if the firefighter had been properly trained for the particular job. The officer in fact did not know the answer to the chief's question, but said that since the firefighter was a veteran of the department, he would certainly assume that the firefighter would know how to do the job. "Never assume anything," the chief told him. "While some people may have had thirty years of experience, others may have had only one experience, repeatedly, for thirty years."

The chief directed the officer to train and work with his company's members to make sure they knew how to perform their job tasks and responsibilities properly. Getting involved in the training re-established the officer as the company leader. He commanded renewed respect as well as admiration and trust. The firefighters were eager to master new skills and to apply them to the job. The company officer now had a solid crew with all the expertise necessary to handle the job at hand successfully and smoothly. Also, the new training program provided a fair basis to use in assessing job performance. If a firefighter did not carry out his on-the-job tasks up to departmental standards, the company officer was able to evaluate the problem more accurately and discipline the firefighter accordingly.

3. **Let them complete their tasks.** Once employees have demonstrated that they have the appropriate skills, get out of their way and allow them to do their jobs.

4. **Let people know that their input is important and that they have control over their participation.** Managers need to be able to delegate responsibilities. In doing so, they must relinquish control over the details of how tasks are to be carried out. Accountability for the success or failure of the project remains with the manager. Proper staff training results in competent workers who can be trusted with decision-making.

5. **Provide people with timely feedback.** The feedback must be objective, honest, and timely. Positive feedback is just as important as corrective feedback.

   Effective leadership should:

   - Make people feel strong and help them feel that they can influence their environment.
   - Build trust in the leader. Officers should be open, honest, and share information on how the company fits into the overall picture of the department.
   - Make sure members know how to do the task before delegating duties.
   - Structure cooperative, rather than competitive relationships.
   - Resolve conflicts by confronting issues together rather than avoiding or forcing a particular solution.
   - Stimulate and promote goal-oriented thinking and behavior.

PROBLEM SOLVING AND MANAGEMENT TECHNIQUES

"Getting People to Want to Do What You Have to Get Them to Do" is the title of a leadership workshop I conducted for several years as a field instructor for the University of Illinois Fire Service Institute. To ensure that I addressed their concerns, I always asked the fire service leaders participating in the workshop to describe one of the problem situations they were currently faced with at work. What follows is a sampling of those situations, along with my recommendations to them based on some of the skill models discussed earlier.

Managing increasing workloads in limited timeframes

A captain told the group that he had more work to do than time to do it in. While his time constraints had remained constant, his workload had steadily increased. When someone suggested that he delegate some of the work, he said he did not have time to train anyone. While acknowledging the initial difficulty of adding the training function to an already unmanageable workload, I emphasized that time spent on training is a worthwhile investment—one that pays off both in the short run and in the long run. By teaching his subordinates how to carry out job tasks that he did not have the time to do, the captain would get the help he so badly needed to meet the demands and objectives of his department. Since his overall workload would be reduced, he would be able to focus more on the managerial and planning aspects of his job.

Another benefit of the training function is that the captain would be able to interact closely with his subordinates in the process. Opportunities for such healthy and positive interaction should not be passed up. Working together toward a common goal helps to foster trust and loyalty between supervisors and subordinates. This type of interaction also gives workers a sense of the importance of their role within the department, and it allows supervisors to give workers the recognition they strive for. When supervisors acknowledge their subordinates' achievements, it helps to fulfill the workers' needs for self-worth and self-esteem. As we discussed in the Communications Model (chapter 10), the need for positive self-esteem is one of the five basic human needs. The satisfaction of this need for recognition results in positive behaviors in the workplace and contributes to high morale.

Providing leadership when leadership at the top is lacking

A lieutenant described the frustration he was feeling because he did not believe that his supervisors possessed the skills necessary to handle their authority. It is not uncommon for subordinates to feel that the bosses are not as smart as they are. However, they are the bosses. In addition to that, they are people, and as human beings they are capable of learning and changing. I asked the lieutenant to identify any particular skills that he believed would help his bosses improve as leaders. I suggested that he put together a program that would help them appear to be more effective leaders. The lieutenant would have to be willing to allow his supervisors to take the credit for the program; his own benefit would simply be in knowing that the common good would be served. If his personal agenda objectives were to promote the common good, then he would be comfortable with this tact.

After discussing the lieutenant's specific case, I asked the workshop members to reflect on why they went through all the work and effort they did to become supervisors. I asked them to think about what the driving force was behind their quest for advancement in the fire service and why they really wanted to be bosses. Albert Einstein said: "Boys/Children want power so they can be somebody. Men/Adults want power so they can serve somebody." Our reasons for desiring power affect the way we use power. People who are trying to "be somebody" tend to create defensiveness in those with whom they interact because people resent being used solely for the purpose of advancing someone else's selfish agenda. On the other hand, people who use power to serve someone else usually reduce tension in the people around them and promote positive attitudes as well as feelings of openness and comfort. The key to getting others to want to do what you need them to do is to tie a personal value for them to your objective. They need to have a stake in the success or failure of the project. This is just as true for your supervisors as it is for your subordinates.

Getting cooperation from and improving communication with newer employees

A battalion chief said his department was having problems with the younger recruits. He went on to say that when they are directed to do something, they always ask why. To the chief, it seemed as if the recruits had a problem following orders and that they questioned his authority. He thought that perhaps the young recruits' lack of military training was at the root of what he considered their inappropriate behavior. Certainly, military veterans would bring a different set of experiences to the job—and probably a different attitude as well—than would younger people with no military experience. I asked the chief to consider for a moment the possibility that the recruits may be asking why something needs to be done in order to feel more comfortable with the task, and not out of insolence or arrogance.

In general, if the people doing the work know the reason why a given task is done in a particular way, they are more likely to feel better about their participation in the project. Is it asking too much of supervisors to tell task providers why something should be done, and how? I don't think so. It is one way of aligning the perceptions of everyone involved, so that all parties approach the project from the same perspective. As I have said many times in my workshops, "Where the *why* is known, any *how* is possible."

The ideal time to provide the answer to any "why" question is during a training session for that procedure or subject. But if a "why" question comes up outside a training session in a non-emergency situation, the question should be answered at the time it is asked. During emergency situations, however, the subordinate should carry out the order first, and then ask the question when the crisis has passed and the supervisor has a reasonable opportunity to answer it.

As I said in the first part of this chapter, training is indeed one of the supervisor's most important functions. Subordinates cannot be expected to take on responsibilities and duties that they have not been fully and professionally prepared for, including subordinates who have been on the job for many years. Even long-term employees may not have been exposed to all areas of a department or had involvement in all aspects of service. So, it is a good practice for supervisors to discuss assignments with their subordinates ahead of time and to assess what kind of training will be required before the tasks can be performed.

When employees are given work to do without adequate instruction, they often feel resentful toward their supervisors for putting them into the position of having to perform a task that they are not adequately prepared to perform. Proper training helps build healthy relationships between subordinates and supervisors, and it ensures that the job at hand will be carried out satisfactorily.

Dealing with system abusers

One of the supervisors in the group explained that he was having a problem with members of his department taking undue advantage of the department's employee benefits program, for example, using excessive sick days. One of the other supervisors attending the workshop reminded us that we all must be careful not to deny any employee the benefits that he or she is rightfully entitled to under the company plan. All employees must be treated equally at all times, and caution must be used when addressing specific issues of benefits entitlements to individual employees.

I suggested to the supervisor that a thorough review of benefit procedures for all applicants might be a good place to start correcting his problem situation. As is true with all regulations and procedures, evenhanded enforcement is essential to the integrity of the system. Fair and consistent enforcement—coupled with fair, well thought out, and clearly written regulations—will usually identify and deal with system abusers.

Addressing apathy within an organization

One of the workshop participants asked for input on how to reverse what seemed to be a trend toward apathy and sluggishness within his department. It is not uncommon for an organization to have these occasional flat spots with low-energy levels. The up and down swings of a company are not unlike those fluctuations we experience as individuals. As we have discussed earlier, when we feel unhappy, out of balance, or stagnated in our personal lives, we must reexamine our personal agenda, and make the changes necessary to recapture and refocus our energy. Similarly, when the energy and motivation of an organization's members are flagging, the leadership must review the organization's mission statement.

Reviewing the mission statement can serve as a clear reminder of why a company's members should work hard at what they do and why the goals they are striving to achieve are worth the effort. Taking a fresh new look at the mission statement can go a long way toward re-igniting the motivation and passion of a company's workers and its leadership.

Sometimes, however, a review of the statement does not inspire renewed commitment, but rather reveals the need for revision. Leaders should look at the organization's stated objectives and then assess what progress has been made and what changes are needed. The mission statement may simply be outdated. The stated goals may indeed have been met already; in which case, feelings of apathy and lack of motivation among workers would not be surprising. The organizational mission statement, like the personal agenda, is subject to change as the reality changes around us. Supervisors can also stimulate renewed interest and enthusiasm throughout the organization by including all members in the revision process. By seeking and welcoming each person's input, supervisors can give subordinates some of the recognition that we all need as human beings. All of us need to be appreciated and given the opportunity to contribute.

A few of the workshop participants told the group that they thought the apathy and low morale within their departments was caused by the lack of any kind of mission statement at all. When individuals have no personal agenda, they have no focal point; when an organization has no mission statement, its members have no standards to hold to and no direction to follow.

The participants who had never been involved in the creation or the revision of a mission statement were unsure how to go about it, so the workshop provided a forum for a full discussion of the process. I recommended that each member of the group first look at the personal agenda guide (appendix A), and then make appropriate modifications in order to develop an organizational mission statement. First, they needed to identify the areas of responsibilities for the organization; and second, they should describe how these responsibilities were to be met. In doing so, each supervisor would have to include as many of their subordinates in the process as possible, making sure that all employees had opportunities to gain some positive recognition. Although the process is open and all employees are encouraged to contribute, the leadership can use the power of veto when conflicts arise or when it is otherwise prudent.

Certainly the creation of a mission statement provides a focal point toward which members of an organization can direct their commitment, energy, and ingenuity. Leadership within the organization has the responsibility to urge the members to create opportunities to fulfill those commitments and bring their ideas to fruition. Because of the different elements that contribute to each situation, however, there is no single magic solution to all apathy problems. The skill models presented in this book, coupled with the effective use of our human endowments, should result in reduced apathy and increased motivation within an organization.

Chapter Fourteen

Embracing Responsibility ~and~ Trust

The Million-Man March

ON OCTOBER 16, 1995, virtually one million African-American men gathered at the steps of the Capitol in Washington, D.C., in one of the most unforgettable demonstrations of peaceful assembly in American history. The fact that nearly all of the demonstrators were Black males made the remarkable event, known as "The Million-Man March", all the more significant.

Throughout this country's history, the Black male has been demeaned and labeled irresponsible, violent, and uncivilized. The Million-Man March was an unequivocal statement that Black men today refute that image. Washington was sent a message that October day that the Black men of America are not only responsible human beings accountable for their actions, but also strong citizens who will not accept less respect and accord for either themselves or the other members of this society.

One speaker after another took the podium to voice his dedication to living up to his responsibilities and fulfilling his potential. The speakers and demonstrators encouraged one another to approach their roles as community members and as fathers, sons, husbands, and brothers with passion and commitment, dignity, and hope.

At the march, Minister Louis Farrakhan, the controversial leader of the Nation of Islam and the event's organizer, quoted from a speech reportedly delivered by British slave owner Captain William Lynch in 1712 on the banks of the James River in Virginia. Lynch, from whom the term lynching derived, urged American slave owners to keep their slaves under control not by hanging them and thereby depleting the labor force, but by teaching them to trust only the masters and their families. According to Lynch, the slaves should be trained to distrust everyone else because distrust is actually far more powerful than trust. To those who followed his plan, Lynch guaranteed loyal slaves and slave descendants for 300 years.

Considered in the context of Sigmund Freud's theory that normal people ought to be able to love and trust, this scheme—as abhorrent and mind-boggling as it is—may very well have been the earliest acknowledgment of slaves as people. Although Freud developed his theory some 200 years after Lynch's speech was supposedly delivered, the theory is nevertheless applicable. By destroying their slaves' ability to trust, slave owners would also destroy their ability to function as normal people.

I am suggesting to you that regardless of our present circumstances, we have the ability to alter the way those circumstances affect us. The effective use of our human endowments is the key to controlling the impact of our current circumstances and the key to shaping our future circumstances as well.

The first and most important step in taking control of our lives is to see ourselves as human—not as black, white, brown, red, or yellow, female or male, or part of any other group that we may fit into simply by incident of birth. None of us had anything to do with which groups we were born into. Furthermore, even though there are factors that undeniably differentiate us from one another, we are all bound together by our shared humanity—the common denominator among us all. And as human beings, we have 100 percent of the responsibility for our actions and our behavior and for using the endowments we were gifted with at birth to create opportunities to meet our responsibilities and to achieve our personal goals.

Trust, like love, must begin within ourselves. In the process of measuring up to our responsibilities and achieving our objectives, we establish a basis for trusting our own abilities. With each lesson learned, we build trust in our ability to learn. Some of us will find learning painful, but we must not allow the pain to diminish the fact that we are learning something. Becoming comfortable with the learning process increases our self-esteem, and feeling good about ourselves provides the basis for feeling good about others. Establishing trust and love within ourselves becomes the beginning of trusting and loving others. The concepts described in these pages work equally well among all human beings regardless of race or gender. To the downtrodden, victimized, and seemingly powerless, there is hope.

When we conduct our affairs in a trusting and trustworthy manner, we go a long way toward crippling the strength of Lynch's theory. Our actions would fly in the face of Lynch's scheme, thwarting his intent and discrediting his racist proclamations. While there is no concrete proof that Lynch actually made the speech, many observers nevertheless contend that some Blacks behave as if they have been conditioned in accordance with principles put forth in it. So that you can fully realize just how repugnant and intolerable these principles are, I have included a copy of the speech in its entirety below (from www.CNN.com):

Gentlemen,

I greet you here on the bank of the James River in the year of our Lord one thousand seven hundred and twelve. First, I shall thank you, the gentlemen of the Colony of Virginia, for bringing me here. I am here to help you solve some of your problems with slaves. Your invitation reached me on my modest plantation in the West Indies where I have experimented with some of the newest and still the oldest methods of controlling slaves. Ancient Rome would envy us if my program were implemented. As our boat sailed south on the James River, named for our illustrious King, whose version of the Bible we cherish, I saw enough to know that your problem is not unique. While Rome used cords of wood as crosses for standing human bodies along its highways in great numbers, you are here using the tree and the rope on occasion.

I caught the whiff of a dead slave hanging from a tree a couple of miles back. You are not only losing a valuable stock by hangings, you are having uprisings, slaves are running away, your crops are sometimes left in the fields too long for maximum profit, you suffer occasional fires, your animals are killed. Gentlemen, you know what your problems are I do not need to elaborate. I am not here to enumerate your problems, I am here to introduce you to a method of solving them. In my bag here, I have a foolproof method for controlling your Black slaves. I guarantee every one of you that if installed correctly it will control the slaves for at least three hundred years. My method is simple. Any member of your family or your overseer can use it. I have outlined a number of differences among the slaves: and I take these differences and make them bigger. I use fear, distrust, and envy for control purposes. These methods have worked on my modest plantation in the West Indies and will work throughout the South. Take this simple little list of differences, and think about them. On top of my list is "age," but it is there only because it starts with an "A"; the second is "color," or shade. There is intelligence, size, sex, size of plantations, status of plantation, attitude of owners, whether the slaves live in the valley, on the hill, East, West, North, South, have fine hair, coarse hair, or are tall or short. Now that you have a list of differences, I shall give you an outline of action — but before that, I shall assure you that distrust is stronger than trust and envy is stronger than adulation, respect, or admiration.

The Black slave after receiving this indoctrination shall carry on and will become self-refueling and self-generating for hundreds of years, maybe thousands. Don't forget you must pitch the old Black male vs. the young Black male, and the young Black male against the old Black male. You must use the dark skin slaves vs. the light skin slaves and the light skin slaves vs. the dark skin slaves. You must use the female vs. the male, and the male vs. the female. You must also have your white servants and overseers distrust all Blacks, but it is necessary that your slaves trust and depend on us. They must love, respect, and trust only us. Gentlemen, these kits are your keys to control. Use them. Have your wives and children use them, never miss an opportunity. If used intensely for one year, the slaves themselves will remain perpetually distrustful.

Thank you, Gentlemen.

Remember, even a broken clock is correct twice a day. Such is the case with Lynch's speech. It identifies misplaced trust and distrust as the keys to maintaining oppression. Appropriate kinds of trust then may very well be the key to breaking the cycle of oppression.

Assessing risk is part of the process of effectively using our human endowments, including who to trust and how much to trust them. It is important that we trust ourselves based on the belief that we can learn what we do not already know in order to fulfill the desires of our heart-of-hearts.

A TRUST MODEL

A good formula for trust is:

Relationship *multiplied by* **Knowledge,** *divided by* **Risk**

Let's examine the three factors in the trust equation:

- **Relationships.** Are our relationships with the other parties involved positive or negative, nourishing or destructive?
- **Knowledge.** Are all parties involved equally knowledgeable? If so, is that knowledge important or valuable to our agendas?
- **Risk.** What risk, if any, is involved?

You can weigh the answers to these questions whenever you need to make a true evaluation of trust levels.

NO-FAULT SOLUTIONS THAT BREAK THROUGH STEREOTYPES

Racial Profiling is an act by which one targets another on the basis of that person's ethnic origin or race. Generally, this type of behavior, especially since the mid to late 90s, has been attributed to law enforcement agencies. According to the American Civil Liberties Union (ACLU), "Racial profiling is prevalent in America." The ACLU goes on to say that, "Today skin color makes you a suspect in America. It makes you more likely to be stopped, more likely to be searched, and more likely to be arrested and imprisoned."

One of the most powerful outcries against the practice of racial stereotyping came to light several years ago, prompting everyone from social critic to politician to agree that *driving while Black* is a real fear for Black people in this country. A reality documented in almost every metropolis in the United States, especially Los Angles, Chicago, New York, and New Jersey. The battle on the streets, highways, and across the country is well documented.

Even wealth and fame do not deter the awful stench of a profiler's expectations. Joe Morgan, hall of fame baseball player, was seized on the premise that "all Black men are suspects". Jamall Wilkes, former Los Angles basketball star, was erroneously stopped and handcuffed. Al Joyner, 1984 Olympic medallist, was stopped twice within 20 minutes, handcuffed, and made to lay spread eagle on the ground at gunpoint. Wesley snipes, a well-known actor, was taken from his car at gunpoint, handcuffed, and forced to lay on the ground while a police officer knelt on his neck and held a gun to his head. Another actor, Blair Underwood, was also stopped and detained at gunpoint. African-American businessmen and professionals report similar treatment. Deval Patrick, the head of the Civil Rights Division of the Justice Department, reports that he still gets stopped if he is driving a nice car in the "wrong" neighborhood. Christopher Darden, the O.J. Simpson prosecutor, says in his book, *In Contempt,* that he is stopped five times a year because he drives a Mercedes-Benz. "I always seem to get pulled over by some cop who is suspicious of a Black man driving a Mercedes."

To profile a person or group, one simply does what most of us in some way or another often do: we play into the trap of stereotyping. As notorious as these revelations became going into the new millennium, the question for us today is: What causes this behavior? Are Blacks and other so-called "high crime" minority groups the only ones victimized by these actions? Is profiling limited to police action? Or, does it play into our everyday living? Each of these questions is easily answered when the root of the virus is more broadly examined. First, we must understand that when we stereotype people and believe that these stereotypes are in some way valid, we begin to have expectations of that person or group. Obviously, if someone is always late, and you expect that individual to be late, you have a preconceived notion of how that person behaves. Is that wrong? Of course not. A repeated pattern of interaction warrants an understanding from the person being affected. Now, if the person who is always late just happens to have blonde hair, and we, based on our interactions with this individual, expect that all people with blonde hair will be late, then we are guilty of developing a stereotype for people with blonde hair. An extension of this is the behavior we exhibit based on our stereotype. For example, we may begin to chastise blondes for behavior that resembles lateness. Like misdirection, we are guilty of profiling. Stereotyping without action is stereotyping, stereotyping with action is profiling.

If this makes sense, then we must agree that everyone is susceptible to stereotyping anyone and everyone. That said, how do we confront the problem? The answer is as simple as the trap that so often ensnares us: people should only be judged by their actions, not by who or what they resemble. If a White male assaults a woman, he is a batterer and should be handled as such. The woman has every right to judge this man based on his actions toward her. But, if the same woman now labels all White males as batterers, she has been twice-victimized, first by the assault, and second, because she has developed a preconceived notion about White males. That notion may forever impair her ability to engage another White male in a healthy manner. From now on, this woman will be apprehensive and guarded around men who have never harmed her. She has now profiled White males, and as such, looks for the same qualities in all White males. She may avoid and blame all men who represent the group that she sees as her victimizers.

This simplistic scenario plays out for any group or individual who does not recognize that stereotyping exists and that it is dangerous and self-limiting. Similar situations simmer and boil over into every part of our lives.

Sometime in 1975, Lieutenant Raymond Hoff called me at the firehouse and asked if I would be interested in joining his staff at the Chicago Fire Department Training Academy. I told him that I was not interested. He then reminded me of a conversation that we had when we were both assigned to Truck Co. 15 several years earlier. The conversation had to do with the advantages of being White and the disadvantages of being Black members of the Chicago Fire Department. I must have mentioned that White members could expect to have opportunities that Black members could not realistically expect. He then said that he was offering me an opportunity that could help my career expectations. He said that he could not promise me anything special and that it might not be easy. The phone call ended with me agreeing to attend a two-week training program conducted by Lincoln College.

Attending the class was the fire commissioner, Robert J. Quinn, and his entire staff. The rest of the class members were hand-picked by the director of training and his supervisors, Ray Hoff and Captain Dave Rice. I was one of three firefighters in the class. It was intimidating to be among so many bosses in a room at the same time. As the class went on, we started to relax with them as we divided into work groups. At the end of the first week, each of us was required to make a short presentation. It was one thing to make a presentation to a room full of supervisors as firefighters, but as Black firefighters, Clarence Ellison and I were terrified. Allen Schlueter, the other firefighter, told us that it was not easy making a presentation to this group as a White firefighter either. We were three scared people. I went to my doctor to get a prescription for a tranquilizer. After I explained the situation to him, he gave me a prescription for Valium. After taking one, I felt numb on one side of my head. That scared me, so I flushed the rest of them down the toilet and decided that I was going to say a prayer and deliver the presentation that I had prepared.

Things went well. During the question and answer session that followed my presentation, one of the chief officers asked me what made me think that White firefighters would pay any attention to anything that I had to say as a Black firefighter. After thinking about it a minute, I realized that this question went to the heart of the White and Black stereotyped roles of White superiority and Black inferiority. I asked if someone brought a gift to him, would he be more interested in the wrapping that it came in than he would the contents of the package?" He agreed that it would be unreasonable for one to be more concerned with the wrapping than with the gift. Continuing, I said, "I would expect their undivided attention because I would give them information that they could use." That was the beginning of some strong friendships that have lasted to this day.

It was at that point that I realized I do not have to be bound by Black stereotypes. It became clear to me that what I think of and expect of myself affects my behavior more than what others think of me. We tend to behave in direct correlation to our self-perception. When we perceive ourselves as powerless, we behave as if we are powerless. True to the parable, "They can because they think they can." See chapter 2 for help designing a personal agenda.

Being an experienced firefighter assigned to Truck Co. 15 for the better part of nine years during the 1960s was a great credential. Truck Co. 15, Engine Co. 45, and Battalion 16 were housed in the same firehouse on the southeast side of Chicago. The record books show that during that time period, we were one of the ten busiest firehouses in the country. Each of these companies made more than 3,000 responses annually. Until 1967, Black firefighters and White supervisors staffed the house. Early in my career, White supervisors who were in some kind of trouble were assigned there as punishment. Later on, in the mid 1960s, White officers requested assignments to the house because they enjoyed working on the busy companies located in the middle of the south side's ghetto. We were a proud bunch who took pride in our firefighting skills.

The Chicago Fire Department was integrated in 1967 following an accident involving a firetruck that struck and killed a Black woman. The unfortunate woman had been standing on the street across from a firehouse on the west side of the city. A race riot erupted and Black firefighters from the south side were sent to the west side, and White firefighters from the west side were reassigned to south side fire companies. Apparatus operators were not moved in this exchange. As a driver of Truck 15, I remained on duty there. This integration, which was thought to be impossible a few years earlier, was accomplished overnight by telephone without incident. As we measured each other's competence levels, and we learned that we could trust one another, things returned to normal. Only the faces were changed; the game was the same.

In a recent conversation with Chief John (Jack) McCastland, Deputy Director of the University of Illinois Fire Service Institute, he recalled being in Vietnam during this time period. He told of sitting with an integrated group of fellow marines, each one reading news reports of the civil unrest back home. After a while, they thought out loud, "What is wrong with us that we can get along, and the people at home cannot?" The reason they were able to live together without conflict was that each one knew that the other one was trustworthy. I am convinced that the solution to breaking stereotypes is for everyone to be held accountable and responsible for his or her behavior. In so doing, unearned superiority and irresponsible behavior will not be tolerated.

When I joined the training staff in 1976 as an instructor, I conducted training for working companies sent to the academy to update their skills and to exchange tips on how to be more effective on the fire scene. In 1977, the first group of recruits hired as a result of affirmative action entered the academy for their basic training. A court decree had directed that 40 to 60 percent of the 150 recruits be minorities. Never in the history of the Chicago Fire Department had there been such a large number of Black and Latino firefighters in one place at one time.

There was a bit of confusion as new instructors were brought in to train. There was a staff meeting of deputy fire commissioners who were very unhappy because the court's decree forced the hiring of so many minorities at one time. Picture this in contrast to the previous hiring that was the traditional 90 to 95 percent White. The shock was evident when one deputy commissioner stated in the meeting that, "They can make us hire them, but they can't make us keep them." He then directed the director of training to develop the most difficult training curriculum possible.

The city had recently signed an agreement with the state fire marshal's office to make state certification a condition of employment. This meant that a curriculum had to be developed to meet the state's criteria. I was given the assignment of serving as a counselor to help both the instructors and the new recruits adjust to one another. This was an excellent opportunity to study stereotypes and watch them change as people came to learn more about each other. In addition to getting to know one another, members had to learn to co-exist and assimilate into a paramilitary organization at the same time.

In some ways I think that the unyielding demands of a paramilitary regimen contributed to the overall success of those efforts. Many of the recruits found that for the first time in their young lives they were being held accountable and responsible for their behavior. Using equally administered progressive discipline that permitted increased penalties for repeat offenses up to and including termination leveled the playing field. Many of the Black recruits were treated as equal to Whites for the first time in their young lives. It was brought to my attention that White members of the department felt that affirmative action would benefit them also because being White without a political sponsor was about the same as being in a minority. Affirmative action required federal government oversight that would assure them equal treatment within the promotional process. I came to understand that when a White firefighter says he is being treated like a minority, he means that he has to work for any respect he gets.

There was an incident that occurred when a group of Black recruits, sensing that both the White instructors and the White recruits were confused and uncomfortable with their presence, started a movement to become an intimidating force to take advantage of this situation. Their intentions seem to have been to do all that they could to continue the confusion and to intimidate the White recruits. They boasted that they were going to run the training academy. Once the training staff was made aware of the situation, all the recruits were asked to report to a large meeting room. They were put on notice that sophomoric behavior would not be tolerated, and any violations of department policies would result in the full force of the discipline procedure being brought to bear on the violators. They were informed that the city of Chicago was committed to providing a safe secure workplace for every employee. Anyone in the room who felt that they could not abide by department policies was asked to leave the room and resign. A short question and answer period clarified the rules that were to govern their behavior. They were then released to their group instructors. The instructors were directed to meet with Captain Raymond Orozco on matters regarding Latino cultural concerns (Captain Orozco retired as fire commissioner) and me regarding Black cultural concerns that might develop within their individual groups.

After that meeting, there were no major problems, only some minor attitude adjustments as some people struggled to change their stereotypical behavior. Most of the adjustments were made by assuring the members that they could expect to be treated as equals and judged by one set of rules. Not unlike Jefferson in *Lessons Before Dying,* as their self-perception improved their behavior improved also. Changing stereotypical behaviors is difficult because we tend to develop a comfort level with them. It is convenient to be excused for your irresponsible behavior because you are Black, or to be recognized for your unearned superiority because you are a White. When we hold people accountable and responsible for their behavior, we help them break out of their stereotypical behaviors.

The following situation took place in the spring of 2001. The town is located in the southwestern part of the United States.

The chief of a southwestern town reported that, one day he had a race problem and didn't know why it happened or what to do about it. Some tools, equipment, and money had been discovered missing. The chief said that the member who was responsible was removed from his position while an investigation was being conducted. That member, who was Black, denied that he had done anything wrong and felt as though he was being picked on because he was Black. Politicians from outside the community came to town to take up his fight against racism in the fire department by placing pickets around the firehouses. The pickets let the emergency responders out of the firehouses, but blocked their entry back into the firehouses when they returned.

Just before these charges, racial relations appeared to be very good. For example, a Black firefighter who had recently lost his father was having trouble getting enough money to bury him. All of the firefighters, Black and White, without hesitation, put together enough money for the burial. Talks with the Black firefighters seemed to go well until they met with the outsiders following our meetings with them. Then, additional subjects of discontent surfaced, and as fast as one issue was settled, they would present another one to the chief and his staff. The White firefighters began to think about reacting to the name-calling and accusations of racism they were receiving. Some of them had longstanding relationships with Black members who were on the other side of this situation. The department became divided along racial lines. We were at a loss as to what we should do.

I suggested that the chief keep things from getting any worse by trying to hold back the White firefighters from retaliating against the verbal attacks. They had done nothing to deserve those attacks. It was also very important that the chief keep all communication channels open. The chief responded that he did not know how much longer he would be able to continue talking about any of that because the Black fire-fighters were threatening legal action. If they had taken legal action, the chief would have to turn everything over to the town's attorneys and step out of it. I asked him how he felt about that possibility. He said that he would welcome it because objective talk in the courts would take the place of the emotional tirades and name-calling going on in the streets. He believed it might reduce the emotional tension. I advised him to keep the mayor and town attorneys informed of the day-to-day events so that they would be able to answer questions that might be asked by the press or politicians.

I also advised him to try to think in terms of doing the right thing by following the dictates of his department's policies. If there were any violations of those policies, he would have to answer for them personally. "Be sure to play by one set of rules," I cautioned. If there were ever any evidence of uneven treatment toward either side, the town would have to answer for it. I added, "Make sure that your house is in order before you go any further along." The chief then said that he would call in outside experienced investigators to conduct a review of his entire department, from his own position all the way down to the newest member of the department. I commended him for that kind of determination and told him that that type of action would quickly bring things to a conclusion. My suspicion was that the Black member involved with the missing inventory was using race as an issue to confuse and hold off the inquiry into his case. That is a ploy that works best when there are multiple agendas at work and when the organization cannot survive a close critical investigation without uncovering additional violations of policies or regulations.

The following excerpts are from an officer empowerment workshop that I conducted with the officers of the department that experienced the race problem.

One of the elements of racial relations I have observed is that very often normally thoughtful, rational, and objective-thinking White people appear to get emotional at the mention of race. Calling a White person a racist to their face often seems to be a traumatic experience for them. I think that the very least they should do is ask the accuser for proof. Asking the accuser how they reached their conclusion would be a good place to start. The chief in this case responded perfectly by starting an investigation to determine the validity of the charges. This is important because in an effort to avoid being called a racist, people sometimes over-accommodate minorities. By doing so, they behave in a racist manner. Treating someone in a certain manner because of their race or gender is a racist or sexist act, even if the treatment is positive.

Let's look at some stereotypical roles that might come into play here. The Black stereotype calls for Blacks to accept roles that will not hold them completely accountable or responsible for their behaviors. Moreover, when they are held accountable and responsible, their role allows them to make excuses. When all else fails, they can claim that the supervisors and the organization are racist. Living this stereotype requires child-like behavior. On the other hand, the White stereotype calls for them to be given recognition for unearned superiority, and as such, they are to be forgiving of the irresponsibility of Blacks. "I may be poor, but at least I am not Black" is a statement that speaks to the White stereotype. A White supervisor was quoted as telling a Black worker, "If you can stay out of the cemetery, I can get you out of jail." Living this stereotype requires parent-like behavior. It goes on and on in a self-defeating cycle. What is the solution to this craziness?

The simple truth is that each human being must accept the responsibility for their behavior. We must all expect to be held accountable and responsible for our choices and the effective use of our imagination, independent-will, and gift of self-awareness. I recommend that Blacks reject their stereotype of irresponsible behavior and the inferior status that comes with it, accept responsibility for the effective use of their God-given endowments, and rise to the common ground of equality. In a like manner, Whites should reject their stereotype of unearned superiority and come to the common ground of equality. The class agreed that these recommendations would be a good place to start working on the resolution of their racial issues.

Several months passed before I got back to the chief to follow up on his racial problem. He told me that there had been an Equal Employment Opportunity Commission complaint filed on behalf of the Black members of the department. However, the complaint was withdrawn for some unknown reason (about the same time that the politicians left town). The chief informed me that the investigation into the missing inventory was continuing, as was the impartial review of his department's operations.

In 1985 as part of a cultural awareness program that I conducted for recruit firefighters and paramedics entering the Chicago Fire Department, I devised a racial and gender stereotyping exercise. The purpose was to determine how members of different races and genders viewed themselves and others. A color was assigned to racial and gender groups instead of names or race to protect the personal identification of the participants. These colors were then placed on forms. Participants were asked to complete the forms by describing each group. When completed, the forms were collected, and the results were reviewed. The responses are listed in the section, "How Do We See One Another?" on page 120.

The discussion that followed revealed that many of the participants were surprised to find that they were viewed much differently than they would have imagined. Even though they could understand why others would view members of their race or gender the way they did, the views did not apply to them individually. It became very clear that group stereotyping was not only inaccurate, but also unfair to individual members of the group.

Are they as human as we are? Are we as human as they are?

The answers to these questions may determine the way that we relate or respond to people of different cultures or backgrounds. The differences may be obvious, but how significant are the differences as they relate to our interacting with them? These questions are not usually asked out loud, but many of us may ponder them as we encounter people that we perceive as different from us. There is some merit to the notion that in order for some human beings to mistreat or abuse another, they must first dehumanize the victim. After that, since the victim is less than human, they are fair game for whatever abuse the superior human being chooses to impose on them. On the other hand, the victim cries out for the respect due him/her as a human being. So, it comes to the questions: Are they as human as we are? And, are we as human as they are? The answer to each of these questions is a resounding "Yes". Read on as I explain.

As I watched a group of workshop students reading the text about "Racial and Gender Profiling", I found it interesting that many of them were smiling. This was interesting because the topic is usually very sobering, and participants tend to become very sensitive and apprehensive. As I began to examine this reaction, it became clear to me that most people had been so involved in developing perceptions of others that they were surprised to find that the others were developing perceptions of them. Their smiles seemed to be asking: "Am I that obvious?" In many instances, they found those perceptions amusing.

HOW DO WE SEE ONE ANOTHER? RESULTS OF THE EXERCISE

In 1985, as part of a cultural awareness program that I conducted for recruit firefighters and paramedics entering the Chicago Fire Department, I devised this racial and gender stereotyping exercise. Responses to the questions below follow this brief description of the procedure that was used.

Purpose of the exercise: Determine how we see ourselves and how others see us. Do we all see each other and ourselves the same way?

Method used: The participants are to answer the following questions:

How do Whites view themselves?
How do Blacks view themselves?
How do Latinos view themselves?
How do women see themselves?
How do Whites view Blacks, Latinos, and women?
How do Blacks view Whites, Latinos, and women?
How do Latinos view Whites, Blacks, and women?
How do women view Blacks, Latinos, and Whites?

What are your pet peeves with Blacks, Whites, Latinos, or women?

System used: In order to protect the identification of the participants, the following system of colors was developed:

Blue = White Males	Red = Women
Green = Black Males	White = TBA
Black = Latinos	Yellow = TBA

Results: After the forms were completed, they were collected and the results reviewed. The discussion that followed revealed that many of the participants were surprised to find that they were viewed much differently than they expected. Even though they could understand why others would view members of their race or gender the way they did, the views did not apply to them personally.

The next section includes the responses we received from the participants.

HOW WE SEE OURSELVES

Blacks on Blacks

Aggressive
Athletic
Ambitious
Rhythmic
Religious
Very proud
Intelligent
Hard workers
Sense of independence
Oppressed, not trusted
Unequal recognition
Have to work twice as hard
Becoming recognized
Sense of achievement
Proud heritage
Added wisdom and strength
Gifted, self assuming, adaptive

Arrogant (quick tempered)
Sense of denial of basic rights
Lost heritage
Stress
Challengers
Street-wise
Physically inclined
Higher cost of doing business
Victims of inferior education
Are discriminated against
Have the ability to adjust to circumstances
Need equality
Desire to achieve
Fair toward others

Whites on Whites

Drive to succeed
Encouraged to work for a better standard of living
Move wherever you like, options
Education is better, "higher"
Higher morals
Family unity
Victims of affirmative action
Fear of racial violence in Chicago
Sunburn easy
Trusted more
Respect for the property of others
Respect for others' religion
Klan plan
Resent minorities
Non-racist
Proud to be an American
White male stereotype
Have to earn everything
We are a target group
Burden of carrying the load, taxes

Reared to be responsible and pressured to succeed
Presumed to be racist
Contributes and respects society
Respects fellow human beings
Suppresses emotions
Self-pride
Self esteem
Family planning
Financial planning
Respect for personal property
We lead the country
Our neighborhoods are the best
Deserve the best
We should get jobs first
Wealth
We are best educated
Blamed for society's inequalities
Reverse discrimination
Have to worry about property
Not organized

Latinos on Latinos

It is great to be Hispanic
Respect wives and mothers
Can deal with all races
Rich culture
Hard working
No confusion on father's day
Good looking
Beautiful country
Minority's minority
Outsiders without a cause
Strive to achieve goals, because we stand out, color and names
Pride, not ashamed of race
Get along with whites and blacks
Strong feeling about right and wrong

Respect others and their way of life
Not treated as equally
Lack of respect from others
Never being able to achieve total acceptance, like to be Japanese for acceptance
Suffer discrimination from all segments of Whites, Blacks, etc.
Close family ties (sensitive)
Very proud
Honorable
I wear the pants
Religious
Ambitious
Doubt there is real opportunity
Hispanic features different to White mainstream

Women on Women

Tries to look her best (looks are important)
Motherhood, child bearing
Trying harder
Image
Has to show more compassion and sensitivity
Must balance family vs. career
Opening opportunities
More aware of capabilities and potential
Guilt and confusion of role, family vs. career
Competition between women
Superwoman
Manipulative
Flexibility in appearance and attire
Maternal responsibilities
Care-taker, custody bias
Survivors, self-sufficient (Black)
More likely to head households (Black)
Treated as 2nd class citizens (Black accepts it)
Talented in the arts (Black)
Religious (older Blacks)

HOW WE SEE OTHERS

Blacks on Whites

Better job
Unity
Feels superior
Better opportunities
Racist
Well informed
Power, leaders
Arrogant
Advantaged, privileged, and egocentric
Target group (male)
Aggressive
Humanitarian
Strong family background (goal oriented)
Political clout
Financial benefits
Family inheritance
Respect
Has little concept of struggle
Closed-minded about certain issues
Reverse discrimination
Traditional
Dominant attitude
More and better choices in life
Secure

Blacks on Latinos

Unity
Aggressive
Pride

Closer to culture
Hard laborer
Isolated (Black, White)
Disadvantaged (jobs)
Togetherness (religion)
Strong family background
Latinos equal to Blacks
Oppression
Education
Work, life style
Traditional
Underpaid, overworked
Religious
Content
Strong willed
Lots of music
Has to deal with language barriers
Lack of civil rights leaders
Stereotyped as illegal aliens
Work four times harder for less money
Good tradesmen
Fair toward others

Blacks on women

Emotional
Sensitive
Dedicated
Self-conscious
Aggressive
Disadvantaged (Blacks)
Privileged (Whites)
Loving
Strong
Intellectual
Family oriented
I am beautiful (blonde hair, blue eyes)

Strong family background
Very protected
Little concept of struggle
Very loyal friends
Organized
Exploited
Discriminated against
Career conscious
PMS
Insecure
Dislike because of color
Disrespectful
Religious

Latinos on Blacks

Stylish
Compassionate to other minorities
Identification with other struggles and causes
Streetwise
Very artistic
O.K. to be Black
Very rhythmic
Have to work much harder
Feelings toward society
Not given the chance to try in society
Feel like they are not getting their fair share
Loss of culture
Looked down upon
They stand up for what they believe in
Have been held back from achieving their goals
When one Black does an injustice all suffer the consequences
Some are prejudiced
Getting jobs on quotas rather than merits
Excuses "If I were not Black it would not be happening to me."

Latinos on Whites (WASP)

Lives off parents
Mamas boys
Better education and jobs
No set traditions to hold them back
More liberal
More accepted in life
Trusted more
Small families
Money orientated
Good surroundings
Don't have to work as hard as minorities
Right clout (political)
More opportunities in society
Proud of American heritage
Are dominant in society
Powerful in government
Thought of as bad guys
Ease of acceptance
Inability to understand what it is like to be discriminated against
Lucky
Some are prejudiced
One bad, all bad

Latinos on women

Uses sex as a tool for opportunity and punishment
More choice in clothing
Very sensitive
Persuasive
Adaptable to changing times
Left behind
Confused on role in society
Has to work much harder to compete in society
Have to be over achiever in the work place
Speak out

Career minded
Think they are capable of doing all man's job

Whites on Blacks

Less pressure for success
Easier to get public, financial, and federal aid
More excuses for failure
Prejudged, whether good or bad
Lack of family structure
Recipient of affirmative action
Don't sun burn easy
Don't have to account for actions
"Nobody respects me."
Are good break dancers
Are treated unequally
Styling in cars and clothes
Today's White man owes us for what their forefathers did
Sexually superiority
Unity
To succeed one must try harder
Must face the Black stereotype
Discrimination in education, employment, and housing
Sports, nightlife, sex, and vacation
Feel discriminated against
Ride more important then crib
Idolize sports figures
Love boom box, dancing, and strutting
Need to succeed for who we are not what color we are
Thanks for TV
Thought of as unintelligent
Untrustworthy
Opportunities cut short
Gang pressure
Pressure to integrate
Pressure to be tough (at least be a good woofer)
Affirmative action

Whites on Latinos

Strong family structure
Good with spray paint
Hard workers
Fear of gangs
Language barriers
No one is going to push me around
Overload car springs (14 per car)
Low paid, hard workers
Healthy eaters (40 ways to cook a bean)
Sports
Lack of opportunity for jobs and political positions
Pride! heritage
It's owed to us or should be given
Everyone should learn my language
Traditional festivals
Night life
To have indigestion
Machismo
Language
Employment
Religion
Clannish
High tempered
Feel more like more of a minority than Blacks
Affirmative action
Don't think that they are a minority

Whites on women

Guys pay on dates
Affirmative action (special considerations)
Easier to show emotions
Manipulative
P.M.S.
Hope I don't break a nail picking up a stretcher

Have to work harder to prove myself
Family oriented
Sexually harassed in order move up the corporate ladder
Favoritism because of sexual gender
Have to dress nice to impress a man
More sensitive
Have more responsibility for rearing children
Traditional female roles
Competition
Financial independence
Family planning
Sexual role
Always in bathroom for hours in pairs
Have need to be equal to males knowing they are capable of certain tasks
F. A.L.N.
Gangs
Litters
Slang

Women on Whites

Macho
The support system of the family structure
Follow in father's footsteps
Male chauvinist
Have more opportunity
Family expectations higher
Sexual
Self-controlled
Double standards
Pressured to be competitive, aggressive, and successful
Supporter
Provider

Women on Blacks

Macho
Desire to do better than their fathers did
Anger
Compete with the White male
Smug, minority quotas
More violent
Weaker family structure
More pride with achievement
Emphasis on material items
Same as whites except for family responsibilities
Sensitive egos
Double standards in sexuality, marriage, and family roles

Women on Latinos

Macho
Environment, culture
Loud in dress and manner
Loud appearance
Defensive
Try harder: caught between Black and White
Macho
Family oriented
Protective
Religious

PET PEEVES

Blacks on Blacks (pet peeves)

Stubborn
Illiterate
Lack of unity or togetherness
Wearing plastic bags, letting poverty beat the odds to succeed
Unable to advance
Black on Black crime
Lack of unity
Won't support Black businesses or each other
Lack of self-control

Blacks on Whites (pet peeves)

Ethnocentric
Stereotypes
Sense of superiority
Sexual fantasies with Black women
Mentality of forefathers
Lack of knowledge of minorities
The way they shake their hair
Nasal accent
Disrespectful, no manners, bitching in public
Racial jokes
They are just a little sick
Racist

Blacks on Latinos (pet peeves)

Bi-racial
Car decorations
Cannot drive

Lazy
Wages
Lack of communication skills
Population boom
Unmotivated
Mexicans want to be White
They ride the Black movement coat tails
They don't contribute to civil rights
The little ornaments in the family Chevy

Blacks on women (pet peeves)

Make up
Untidy
Independence (Black)
Attitude (White)
Some have superiority complex
Disrespectful
Over reactive

Latinos on Latinos (pet peeves)

Gang bangers
Too macho, sexism
They clash in clothes (high heel sneakers)
Too religious
Not trying to adapt to society
Welfare
Arrogant
Do not put forth best effort for self-betterment

Latinos on women (pet peeves)

Teasers
Never satisfied
Competing with the male

Trying to be macho
They use emotions as a weapon
I can do anything I want or get an attitude
Use gender to manipulate

Whites on Blacks (pet peeves)

Financially irresponsible
Black pride/White racism
Lazy
Think that things should be given them
Persecution complex because of past history
Lack of speech
Employment problems
Poor family planning
Want equal opportunity when not equally qualified
Want special treatment
Can't understand their speech
Not willing to work as hard to achieve goal
Know that they will get special treatment
Gangs
Jive talk
Dreadlocks
Loud colors
Oily hair
Exaggerate
Walk
Sleep too much
Too quick to claim discrimination

Whites on Latinos (pet peeves)

Learn English
Gang involvement
Double parking
Refuse to Americanize

Whites on women (pet peeves)

Nagging
Whinnying to women's liberation
Declining femininity
OTR
Tease
Emotional instability
Sexual instigators, manipulators
Expect male to provide
Paid less for same job
Paid same for less job (fire fighters)
Physically weak
Mentally strong
Capable of handling more stress

Whites on Whites (pet peeves)

Feelings of superiority
Know it alls
Too nice on previous sheets
Always complaining
Bleeding heart
Liberals
People with unsubstantiated prejudices
Lack of unity

Women on Blacks (pet peeves)

Lazy, expecting to get something for nothing
Prejudiced
Less respect for women
Double standards
Relationships for status, curiosity, ego
Lack of courtesy and respect

Women on women (pet peeves)

Gossip
Behavior
Sexually manipulative
"Catty"
Try to intimidate others (Black)
Irritable
Too tolerant of negative treatment (Black)
Think that they should be treated special for expectations (Black)

Women on Latinos (pet peeves)

Language barrier
Cat calling
Won't speak English
Emotional

Women on Whites (pet peeves)

Superior attitude towards others
Take women for granted
Purely platonic relationships are rare
Double standards

Chapter Fifteen

CONCLUSION

THE PERSONAL AGENDA chapter of this book should be used as a guide to identify your innermost desires, which then become your life's goals. The Human Behavior Model should provide some insight on how we tend to function as human beings. This insight is important to our understanding of the dynamics of human behavior, which in turn should result in more effective relationships. The Conflict Resolution chapter should help you understand that the major problem in most conflicts is attitude, or emotion, and we can choose not to participate at that level. We can also reduce the time spent dealing with others emoting by using the specific problem-solving terms mentioned in the Communications Model: *who*, *what*, *when*, *where*, and *why*. Once the dialog turns to the facts at hand, the parties should be able to at least agree to disagree.

The Behavior/Attitude/Value Model makes the point that while we may think and feel any way we want without scrutiny, our behavior is subject to reward or punishment. We must all understand that irresponsible, unacceptable behavior produces consequences. It is also irresponsible for one person to protect another person from those consequences. The motivational steps presented here describe methods I have employed successfully in my fire service commands. These steps can be applied in most situations that require working with others.

Each of us has the responsibility to use our endowments to protect our interests and to accomplish our goals without causing unnecessary harm to others. No matter what our station in life, we are all charged with using our gifts effectively. It is in this sense that our humanity is our common ground.

It is my belief that the concepts discussed throughout this book can be useful in reducing the harmful effects caused by the irresponsible behavior of some on the responsible members of our society. As responsible people apply these concepts in their daily lives, many of the irresponsible people they come into contact with will be compelled to behave in a responsible manner as well. That irresponsible family member (and every family has one) will be shown that it is his responsibility to manage his own affairs. As we give those irresponsible family members the opportunity to create their own personal agendas and to use their endowments to fulfill their personal objectives, it is important for them to understand that irresponsible behavior has consequences. They must recognize that they have a choice: they can reap the rewards of responsible behavior, or they can pay the price for irresponsible behavior.

Very often the toughest part of doing what you want to do is actually identifying what it is that you want to do. To help jumpstart the process of identifying your objectives, I have extracted the personal agenda section as a stand-alone workbook. This way people who want to get on with the business of identifying the desires of their heart and laying out a plan to fulfill them can do so without going through the text. Also, counselors, teachers, group leaders, and seminar leaders can obtain additional copies for their clients or group participants. Families who are having difficulties with loved ones behaving irresponsibly will find this stand-alone workbook valuable in relating to those loved ones. I have found that the workbook is most effective when you just present it to the family members and ask them to follow the written instructions. Be sure to stress, however, that the information they record is confidential and that they are not expected to share it unless they want to. It is to be their personal agenda; it is not subject to outside judgment or evaluation. (Some family members seem surprised to have choices and to have to be responsible for those choices.) After giving them the workbook, you may experience a feeling of relief knowing that you have given your loved ones a guide to self-determination and responsibility. The rest is up to them.

Thirty-four years with the Chicago Fire Department, fifteen spent supervising personnel engaged in emergency operations, and raising 3 children to adulthood during forty-two years of marriage has taught me that one of the reasons people don't do better is that they simply don't know how to do better. This book along with its companion, the Personal Motivational Workbook, is intended to give people some direction to help them move toward behaving responsibly. If you have not already done so, open up the workbook, follow the instructions, and create your personal agenda. The results you can expect are an increased sense of self-respect, self-fulfillment, and personal empowerment. These rewards will provide you a solid foundation for valuing cultural differences. Most importantly, you will be doing what you really want to be doing: being driven by the most powerful force in your world—you.

Appendix

PERSONAL
MOTIVATIONAL
WORKBOOK

Based on the Text of

Humanity: Our Common Ground
Your Guide to Thriving in a Diverse Society

(©2000, Bennie L. Crane and Julian L. Williams, Ph.D.)

PERSONAL AGENDA
HUMAN ENDOWMENTS

WHETHER A CONTAINER is viewed as half-full or half-empty depends on the intentions, or the objectives, of the viewer. Those individuals who have assumed the responsibility of emptying the container may posit that the container is half-empty. Contrastingly, those individuals, who take the opposite position, filling the container, may see it as half full. These different perspectives are based simply on those things that are important to the individual. Change the objective and you change the perspective. Generally, when people look at the same situation and come away with different meanings, the differences are based on one's objectives, or agendas, and usually those with similar agendas will come away with similar interpretations.

Typically, one expects to find agendas at professional and organizational meetings. Few of us have undertaken the task of developing a personal agenda. Subsequently, if one fails to develop a personal agenda, one must eventually assume the agenda that society assigns. These individuals will inevitably find themselves following someone else's agenda and haphazardly meeting some of their own objectives. One's agenda should provide an opportunity to identify people and individuals who are important. Your personal agenda should identify the principles that will guide your behavior as well as identify your roles with the people who are important to you. Hopefully, this information will be used as a guide for establishing your goals. After developing a plan of action, your behavior will be based on your agenda. This agenda will keep external behavior in accord with internal objectives, creating a sense of internal harmony, as opposed to moving through life reacting to someone else's program. Eventually, as a result of implementing your own personal agenda, you will become proactive. And, after completing this course of action, you will be in a position to move forward seeking out opportunities to fulfill your stated goals in life. If you are still a little confused, do not panic. Try this exercise:

1. *Sit down with paper and pen.* Use your imagination to create a scene five to ten years into the future. If that scenario seems impractical, set the scene for next week or next month. Plan to invite about five or six of the most important people in your life over for a meal. Make this a testimonial dinner for yourself.

2. *Next, write down their names and their relationship to you.* Write out their testimonies to you. Write down things that you would like each one of them to say about you. If there are six guests, there should be six statements.

3. *From this point, identify the principles that you would like to guide your behavior.* These should be timeless and changeless concepts that are true and important. Further, add these principles to the other statements. Subsequently, you now have the main parts of your personal agenda.

4. *Develop the principles and statements into a narrative statement.*

Now, go forth and live your life in such a way that you will be worthy of the testaments given at your testimonial dinner. Use your gift of self-awareness to identify those things that you can do immediately. Some things will require more preparation than others. Now you have a solid base from which to make future choices. Remember, every person possesses gifts from God that produce choices that often seem complex. Considered within the context of our personal agenda, these choices become straightforward.

Use your gift of independent will to make your decisions based on whether or not they can help you move toward your stated goals. You can then go into the larger community looking for opportunities that meet your objectives. Instead of your behavior being a reaction to another's agenda, use your gift of universal freedom of choice to respond to others based upon your principles and objectives. Your gift of conscience should be used continually throughout this process. I do not intend to imply that your gifts should be used in any particular order; essentially, they are yours to use as you see fit. This demonstrates clearly how you empower yourself.

Prisoners of war have described how they used their imagination and their freedom of choice to reduce the effects of the terrible treatment they received at the hands of their captors. Although the movement and behavior of captives is restricted, the gift of imagination is not subject to outside interference. So, the choices are to either bemoan imprisonment, or take advantage of the time to prepare for a better time to come.

One story is that a captive, using his gift of imagination, designed his dream house in his head. Upon his return to his hometown, he built his dream house and turned an otherwise negative experience into a positive one. Similarly, enslaved Blacks working in the fields sang spirituals to help ease the effects of their harsh lifestyle. It is important to understand that others may restrict your movements and your liberties, but no one can control your freedom of choice. This example will further illustrate my contention.

In the segregated southern part of the United States, an angry band of White men came to the home a Black family and demanded that the father send one of his sons outside to be punished. The son had been seen damaging a watermelon patch. The father informed the leader of the assembled group that he understood the charges and agreed that the son should be punished. He added that he would discipline his son himself. The group accepted the father's word and went away.

This is the kind of transaction that can take place after people empower themselves. The internal harmony that comes with personal empowerment allows one to circumvent, as well as, rise above racial and gender concerns. The story of the father and the group represents a transaction between responsible human beings. The father, although oppressed, was known to conduct his affairs in a responsible manner. Self-empowerment is the ability to choose and manage change. It allows one's life to be self-directed.

THE HUMAN ENDOWMENTS

Human endowments have been identified as gifts that come with birth. They are self-awareness, imagination, independent will, and freedom of choice. Conscience, while not present at birth, is listed under human endowments because it is an important part of the internal harmony that comes with personal empowerment. We cannot have internal harmony without a clear conscience. Conscience is a learned trait. It is developed and learned from authority figures. These endowments are tools that all human beings have to meet their personal objectives, and to protect their interests. Here are the definitions of the human endowments:

- **Self-awareness** is one's ability to evaluate their circumstance and to determine what changes are needed, if any.
- **Imagination** is the ability to create images in our minds beyond our present reality.
- **Independent will** is the ability to act based on our self-awareness, free of all other influences.
- **Freedom of choice** is the freedom to choose the response to any event, or stimulus, that comes from the outside. One may choose not to respond to an event.

- **Conscience** is a deep, inner awareness of right and wrong. It is a sense of the harmony between our thoughts, our actions, and the principles that govern our behavior.

THE PERSONAL AGENDA OUTLINE WORKSHEET

My responsibilities for myself include the following: (1) Maintaining my physical and mental well being, (2) identifying and fulfilling the desires of my heart in order to achieve personal happiness, (3) behaving in accord with my highest held values, and (4) being accountable for my behavior at all times. I am to use my human endowments as tools to meet my responsibilities. While using this guide, keep my rule of engagement in mind: Protect your interest at all times without doing unnecessary damage to the other party.

I perform specific activities to ensure my success to meet these responsibilities. These include diet, meditation, prayer, and exercise.

Follow the instructions on the worksheet and when you have completed this guide, you will have developed a major part of your personal agenda. Please note that five guests are the maximum number you are to invite. Your guests may be individuals representing themselves or a group of people. You may want to identify a spokesperson for a group of people. Groups could include: family members, coworkers, parents, children, aunts, supervisors, or subordinates. It is permissible to have a statement from a group of people.

The following worksheet is designed to assist you in developing your personal agenda. You may invite any number of guests up to five. Use additional paper as needed.

Imagine that you are given the assignment to plan an award dinner for yourself. You are to name the five most important people in your life. They will give a speech and talk about you. Write their names here:

1. ______________________________
2. ______________________________
3. ______________________________
4. ______________________________
5. ______________________________

Next, write down those things you would like them to say about you.

1. ____
____
____
____

2. ____
____
____
____

3. ____

____

____

____

4. ____

____

____

____

5. ____

____

____

____

Combine these statements into a single narrative and you will have the objectives of your personal agenda, developed for and by you from your personal values. It is important that you develop your agenda from your personal values because this allows your outside behavior to be in accord with your internal desires. Some of us feel that this contributes to our personal comfort and well-being. When we behave in concert with our innermost beliefs, we tend to be at ease with ourselves. We are happy when we do what we really want to do and it turns out well. It is a self-fulfilling experience. Conversely, the absence of such harmony between our outside behavior and our internal beliefs results in us being uncomfortable with ourselves. Many people experience depression and become despondent with the sight of themselves in the mirror.

This agenda should serve as the guide for your behavior both personally and professionally. You are now to seek out opportunities to behave in ways that will make those statements true. Each one of us has the ability and the responsibility to control our behavior. When the people who are important to us say good things about us, we feel good about ourselves. You may plan an award dinner for yourself as often you like. You have complete editorial rights over it. You can change or adjust it as you see fit. Life is in a state of constant change. So it would follow that one's agenda should change also. Your agenda should be a source of motivation for you. Here is my personal agenda developed using this guide:

Be supportive of others. Convey to others that life is worth living, sharing, and believe that there are answers to most problems. As I help others be all they can be, I become all that I can be. Try to serve, or at least recognize the highest good in others and myself. Worship God by serving others in his honor. Maintain my health with a sensible exercise program and diet.

As simple as it may sound to others, it is motivating to me. When I lose sight of what is important to me, it is good to have my agenda to use as a touchstone. The people that I identified as being most important to me are my mother, wife, and three children. The statements that I wrote for each of them, including my own statement, were the six testimonies that I used to create my agenda. It was interesting that my agenda also serves me well in relationships outside this target group. The sense of well-being and increased comfort within myself transfers into a more relaxed level of participation in my relationships with others.

Review of the Process

- Plan your testimonial dinner to be held next week, next month, or 5 or 10 years from now.
- Identify the speakers you would like to speak on your behalf and the role you have in each of their lives.
- Write their speeches for them. Write out what you would like them to say about you.
- Write out the principles you want to guide your life or behavior.
- Seek out opportunities to make the content of those speeches true by using the principles you have chosen to guide your life.

The Test

Each time you read your personal agenda, you should be inspired to continue moving on, looking for opportunities to meet the objectives of your personal agenda.

INDEX

A

B

C

G

H

I

J

K

L

M

Q

R

S

T

U

V

W–Z